6-4-79

Pathway to Energy Sufficiency
The 2050 Study

Principal Friends of the Earth offices:

124 Spear Street
San Francisco, California 94105

620 C St. SE
Washington, DC 20003

72 Jane Street
New York, New York 10014

Pathway to Energy Sufficiency
The 2050 Study

By John S. Steinhart, Mark E. Hanson,
  Carel C. DeWinkel, Robin W. Gates,
  Kathleen Briody Lipp, Mark Thornsjo,
  Stanley J. Kabala

Foreword by David R. Brower
Edited and designed by Sidney Hollister

Friends of the Earth
San Francisco

Pathway to Energy Sufficiency was originally published, in a somewhat different form, under the title, *A Low Energy Scenario for the United States: 1975-2050* (Madison; University of Wisconsin-Madison, Institute for Environmental Studies, 1977); and appeared, also in a different form, in *Perspectives on Energy,* edited by L.C. Rudiselli and M.W. Firebaugh (New York; Oxford University Press, 1975).

Library of Congress Cataloging in Publication Data
Pathway to energy sufficiency.
   Bibliography: p.
   1. Energy policy—United States. 2. Energy conservation—United States.
3. Economic forecasting—United States. I. Steinhart, John S.
HD9502.U52P37   333.7   78-74807
ISBN   0-913890-31-6

With special thanks to Dharma Press (Emeryville, Calif.), Jeroboam, Inc. (San Francisco), and Nancy Austin.

Cover design by Zenji Funabashi

2061049

Contents

First Considerations

Photo by Tom Turner

Foreword

Like an individual overdependent on alcohol, a society that has over-indulged on oil has a difficult transition to face. The news that the transition is essential is bad news to many, and the bearer of bad news, as in old times, is at risk.

When Amory Lovins suggested in 1972 that the United States could get by with one-third less energy by the year 2000 than the old Atomic Energy Commission was predicting, most energy purveyors laughed. When four years later his *Soft Energy Paths* went on to show that the U.S. could prosper on still less and get to 2025 without nuclear energy and break the fossil-fuel habit by then, even E. F. Hutton could begin to listen. Mr. Lovins showed how thinking people could appraise what their real future energy needs were, match the source to the use, and fare well. His subsequent response to doubters fills two volumes of Congressional hearing record (summarized in *The Energy Controversy: Soft Path Questions and Answers*, Friends of the Earth, 1979). He has elsewhere shown the evolution of U.S. energy needs by the year 2000 in a table that looks about like this in units of Quads per year (quadrillion Btu), the current rate being about 75 Quads:

Year of forecast	Beyond the pale	Heresy	Conventional wisdom	Superstition
1972	125	140	160	190
	(Lovins)	(Sierra)	(AEC)	(FPC)
1974	100	124	140	160
	(Ford zeg)	(Ford tf)	(ERDA)	(EEI)
1976	75	89–95	124	140
	(Lovins)	(VonHippel)	(ERDA)	(EEI)
1977–78	33	67–77	96–101	124
	(Steinhart)	(NAS I II)	(NAS III, AW)	(Lapp)

Abbreviations: *Sierra* (Club), *AEC* Atomic Energy Commission, *FPC* Federal Power Commission, *Ford zeg* Ford Foundation zero energy growth scenario, *tf* technical fix scenario, *Von Hippel* and Williams, *NAS I II III* the spread of the National Academy of Science Committee on Nuclear and Alternative Energy Systems (CONAES), *AW* Alvin Weinberg study done at the Institute for Energy Analysis, Oak Ridge.

]4[

This book focuses on the forecast in the lower left corner of the table—a number one-third lower than a 50-Quad future (Laura Nader's) so low that CONAES apparently chose to dismiss it.

Public thinking changes slowly unless stimulated to do otherwise. We hope people in the U.S. and elsewhere will be stimulated by what Professor Steinhart and colleagues perceive here, and find a new pathway faster than some of us have. I myself remember looking to nuclear power in the fifties as an escape from dams being proposed to provide the ever-increasing energy being sought. Learning of nuclear's many prohibitive costs and knowing we would soon run out of oil, I switched to coal. Watching how they stripmined coal in the Navajo country, its fertility thus destroyed until after a new glacial epoch, I abandoned my hopes for coal, which was not very compatible with clear air anyway. Biomass—solar energy stored by living things—became tempting, especially biomass that was surely going to be wasted. But caution is advisable. Too little is known about when an overdraft at the biomass account might become irreversibly dangerous. We overshot our ability to cope with nuclear side-effects, such as the creation of unmanageable waste and the erosion of truth. We could also exceed our senses by burning fossil fuels too rapidly, exacerbating the carbon dioxide shift now being caused by the destruction of forests, thus altering global climate disastrously. Surely we would not wish to divert too much of the biomass-energy cycle life itself depends upon.

Where to go, then, but directly to the sun? After all, the solar energy sent us free every fortnight about equals all the fossil-fuel energy we can ever recover from the earth. The sun has other errands, however, than immediate human convenience. One-third of our solar largesse must bounce back into space if we want the earth to remain cooler than Venus. Another third must lift water if we want clouds. Of the remaining third, very little is stored by photosynthesis, a process we do not understand yet, to be recycled by still other biological reactions we do not understand either. Moreover, we don't quite want a sky full of windmills, or solar panels where all the most pleasant open space is. So the wise course seems to be the conservative one.

But is there, then, an acceptable conservative course? Probably so: a new look at growth. What can we live with, and what will end us? A society dependent upon using up its nonrenewable resources at an ever-increasing rate will achieve that rate at the cost of its life. Mistaking growth of that kind for progress of any kind is a fatal error, and turning off hearing aids at that kind of news will not make the news any better. *Global 2000*, a federal study in press, bears this out. In every sphere of activity examined, continuing our present course is a pathway to disaster. We can hope that from the United Nations Conference on Science and Technology in Development (UNCSTD) in Vienna, and from the *Global 2000* estimate, enough will have been learned to encourage adherence to ecological law and order as science and technology are applied to development.

One provision of ecological law is that you will last longer if you depend more upon resources that flow and less on those that are stored. If you dip water from a stream, you won't run out; dip it from a pail and you do. Hence, more sun; less coal and oil.

Another provision is that stored resources last longest if made into durable objects designed for recycling. A big car requires in its building the equivalent of 17,000 kilowatt hours of energy, and only another 900 to double its life. The 17,000 go largely to the dull work of running machines, the 900 to using hands and head in a challenging way—not so good for cogs, but better for people.

A third tenet of ecological law is that human intelligence does best when it concentrates on loving the human race more and the rat race less. People who love the human race will sacrifice some present convenience to assure that children have a fair future. They will treat the earth and its resources as conscientiously as a bank does a trust, for reasons fully as good.

And they may find some change in customs beneficial, and not a deprivation at all. The change could provide a chance to climb out of the rut and look around. If one then wishes to drop back in, at least he or she will know it is a rut.

Wanting to get out of a rut myself, as 1976 began, I aimed at the Himalaya, a full fifty years after I first dreamed of going to the highest mountains of all. Getting back into some kind of condition seemed a good idea, and that demanded some changes in lifestyle—less food, less drink, less driving, more walking, better health—changes that saved me three hundred dollars a month for ten months. Twenty-five excess pounds left me in that period, and another twenty-two in the course of a five-week trek and a climb to 18,000 feet elevation. Back home, I could once again wear clothes that had not fit for twenty-one years, and I felt fit enough to try for another sixty-four years. Changes in lifestyle had brought unmitigated delight, and a clear view of the top of the world, too.

Fitness saves energy, especially if you get there by generating your own. James K. Page, Jr. has calculated (in *Smithsonian*, January 1979) that if the rest of the one hundred ten million overweight Americans got back in shape, the energy saved would run the lights and appliances in Boston, Chicago, San Francisco, and Washington, D.C. Or the overfed, by going light with their appetites, could cheer the lives of the underfed and please everyone. And if the overweight American millions all drove five thousand miles less, and walked five hundred miles more, as I did, they would save far more than enough energy to last them all the added years of their lives. More precisely, they would save in fifteen years the equivalent of the proved oil reserve at Alaska's Prudhoe Bay.

Or write your own scenario and tell us about it.

Professor Steinhart and his team have suggested some changes in the style of life and the look of cities and countryside that are much easier to imagine than my scenario, especially by people willing to shelve a few preconceptions in a way suggested by Professor Mark Christensen, of the University of California's Energy and Resources Program in Berkeley. He notes how fast institutions can change when they need to: "Consider the growth of Hewlett-Packard or Xerox; General Motors as an organization was being invented in the 1920s. In 1940 almost no one spoke about independence for European colonies, yet within twenty years the overt political institutions of colonialism had virtually disappeared across the globe. . . . Political and institutional arrangements that make possible new modes of development can lead to extraordinarily rapid and large-scale changes."

The benefits that will accrue to a society that accepts the low-energy scenario could be most attractive. It could take the rush away from the search for further energy—the rush that has been so costly in the past, that has led to unneeded dams, avoidable oil spills, premature draining of Alaska and drilling of the continental shelf, wanton waste of energy resulting from the dismembering of cities and the paving and overdosing of fertile land, the spoiling of waters, the coloring of air. It could spare us further nuclear risk, and lessen the danger of the ugly fight for the residue of resources at the bottom of the barrel.

If people want a world in which people restrain their numbers and appetites, people can achieve it—and will probably prefer it to the grim alternative waiting in ambush unless they do achieve it.

If there is truth in these assumptions, and we think there is, people will perceive it soon. Soon *enough*, we hope, thanks to this book's showing us that altering past trends can lead to a better lifestyle for people and other live things too.

DAVID R. BROWER, President
Friends of the Earth

San Francisco
January 8, 1979

1. Introduction

THE SCENARIO outlined in this book is an estimate of the lowest energy-use future that the authors consider plausible. We have determined plausibility by eliminating those possible futures that require technological miracles, dramatic changes in human behavior, or complete restructuring of national or world political groupings. Our results indicate that a sixty-four percent reduction from 1975 levels of energy use per capita in the United States can be achieved by a few decades into the twenty-first century.

We have attempted to incorporate and extrapolate demographic and settlement relocation trends and to identify some other private trends that might be usefully encouraged. Many of the legislative measures we discuss have been proposed at state or federal levels already, though some may still be far from enactment. Changes of lifestyle implicit in this scenario blend present trends with subjective estimates of how far they might develop.

We have tried to point out some of the economic and employment implications of the scenario, but have not attempted to do any quantitative economic analysis because such analysis is very sensitive to projections of future discount rates and future fuel-price increases. (The discount rate may be seen as the underlying rate of interest in the economy, corrected for inflation.) Published estimates of these two variables differ sufficiently to produce both positive and negative discount rates for energy costs (rates of deflated energy-cost increase minus discount rate). To resolve these differences would require at least some agreement about the quantitative effect of energy-price increases on general inflationary pressures, a relationship about which there is currently much dispute.

We do not propose this scenario as a plan, but offer it as one of a growing number of possibilities in the hope that it will enlarge the sphere of public discussion.

Any normative scenario of the future assumes a backdrop of world and domestic conditions as a context in which the scenario's plausibility may be tested. No scenario of modest size, however, can hope to describe even the complexity of the present, let alone the complexities of an inherently unknowable future. Consequently, a scenario leaves out more future conditions than it describes. Although there is no satisfactory escape from this dilemma, the reader may find useful some views of the future that are in agreement with this scenario.

The resource pressure outlined by Forrester (1971), Meadows (1972), and Mesarovich, Pestel, and their co-workers (1974) seems to us to be more or less correct. Critics of the several reports of the Club of Rome have raised valid criticisms of some aspects of this work, but no counter models exist and the basic results still stand. The projections for the United States and the world offered by Watt (1974), Heilbronner (1974), Lovins (1976), and Stavrianos (1976), and by the Latin American World Model (Herrera et al., 1976), are very close to our own view of the problems we face. We invite the reader to pay special attention to the difficulties of managing ever-larger, ever-more-complex societal and technological systems. In this case, our views correspond most closely with those of Vacca (1974).

The problem is to invent the future. Only the most insistent determinists and some mystical and religious groups demand a single, unavoidable future. In the enormous gap between the unavoidable and the miraculous, wander politicians and academics, ordinary people and fortunetellers, all in their own way trying to estimate something useful about the future in order to reconcile dreams with possibilities.

The general agreement on objectives that characterized the industrial world at the end of World War II vanished as goals were met or abandoned, usually with little or no comment from the three-fourths of the world that was poor. Now there are many new voices in the world, some of them commanding both money and raw materials.

Thus, the future has to be invented in the face of many conflicting objectives—about which there is little agreement—and in a world atmosphere of constant political instability. The process is made even more worrisome by the observable tendency of the industrialized nations to support (or even bring to power) strong central governments for poor nations, even if (one hopes not because) these governments are highly authoritarian, oppressive, and nonrepresentational.

The energy difficulties of the early 1970s have generated close examination of the kinds and quantity of energy and raw materials available in the developed world. Simultaneously, joint problems of population growth and food supply in the poor nations of the world have provided new visions of the future, most of them conflict-ridden or downright disastrous. Response to these visions has been very strange. From the economists, in their role as the leading academic contributors to policy discussion, have come rejections but no comprehensive models that show a less destructive future. Calling for more investment in the less-developed nations of the world, many economists show a faith in technological innovation that is far less frequently advanced by other scientists and engineers.

In a world of growing complexity, there is a greater premium than ever on a correct estimation of the future. Nearly every major nation has groups under official sanction producing forecasts, evaluations, policy impact projections, and other types of assessments. What characterizes these efforts above everything else is their operation within the limits set by their sponsors. Under these circumstances, it is not surprising to find that the results seldom challenge any existing institutions. Yet these analyses, perceived as value free and hard-headed, are often given special weight. Unfortunately, to acquire such a reputation, the analyses are usually restricted to economic assessment, the more obvious deleterious environmental effects, and measures to mitigate the problems. Despite inevitable caveats about social effects or political problems, these analyses always purport to offer the best (or the least bad) policy recommendations. Restricted by their methodology and the nature of their charge, however, the range of options considered by these analyses is very limited. Analyses by independent groups or individuals are often far less optimistic than official projections, but in the competition for credibility may well sacrifice their strongest arguments in efforts to sound unbiased.

The object of this scenario, which lies somewhere between a forecast and a fantasy, is to explore the possibilities of a particular plausible future. We chose a low-energy scenario because it provides the least expensive energy options, but upon further examination we found that it also has some other interesting and possibly desirable properties. For example, the environmental effects of intensive energy use—air pollution, water pollution, land degradation from mining and sprawl, inefficient and occasionally excessive production—are all reduced in proportion to our ability to use less energy more effectively.

There are also international implications inherent in this scenario. First, a real reduction in energy use by the industrialized nations may offer the only way to avoid direct confrontation between the rich few and the hopes of the destitute many; and the United States is the leading user of energy. Second, because of sheer economic power and considerable domestic resources, the United States could play an exemplary role—as some United States politicians believe we now do. Third, the social conflicts and changes in the United States since the mid-1960s make it possible now—in a way that was not possible twenty years ago—for different coalitions and ideals to dominate United States policy.

This scenario incorporates several recent social trends on the theory that it is easiest to persuade people to do something that they already intend to do or are resigned to doing. For example, the U.S. Department of Agriculture notes that in 1976 half of all households in the United States attempted to grow some of their own food. Even

Robert Burroughs, Jeroboam, Inc.

Photo by Robert Burroughs, Jeroboam, Inc.

though the food system requires 16 percent of all energy used in the United States (Booz-Allen Associates, 1976), no one seems to have considered the possible implications of this major social shift for energy policy.

Finally, the idea of central control appears to be weakening both in the United States, where many new local and regional disputes and power groups have appeared, and in the world, where the number of independent voices has steadily increased since World War II. This combination of nationalism at the international level and localism within nations might offer new promise for a world with less armed conflict (though with more rhetorical shouting).

2. Overview

I F FOLLOWED, the scenario described in this book would result in an energy use per capita in the year 2050 of 13×10^{10} joules, which is 36 percent of the 1975 United States level of 35×10^{10} joules. (A joule is a small unit of energy. Humans are chemical engines running efficiently at 100 to 150 joules per second.) To accomplish this end, major changes would be required, but they would not lead to a fall in the standard of living. This contrasts with the frequent assertion that further improvement or continued maintenance of the standard of living is inexorably linked to a constant increase in energy use. Dennis Hayes (1976, p. 6) frames the issue this way:

> Curtailment means giving up automobiles; conservation means trading in a seven-mile-per-gallon status symbol for a 40 mile-per-gallon commuter vehicle. Curtailment means a cold house; conservation means a well-insulated house with an efficient heating system.

Although this scenario goes considerably further than Hayes in reducing the use of energy, his distinction is useful.

Our analysis here takes the traditional sectoral breakdown—residential, commercial, transportation, agricultural, and industrial—and examines each one for possible energy savings. Borrowing from the extensive literature in the field, we describe the posited changes for each sector and aggregate them into an overall primary energy-use estimate, as shown in Table 1.

To bring about these reductions in energy use, changes would have to occur in social and physical institutions and arrangements, not as the result of any master plan but as a consequence of the evolution of tastes, preferences, and organizations. New community patterns, already emerging in demographic trends, are outlined in chapter 3. Shifts in the mix of goods, services, and employment are described in

TABLE 1. OVERVIEW OF LOW ENERGY SCENARIO FINAL ENERGY USE 1975–2050

	1975 Primary* Energy Per Capita 10^{10} Joules	2050 Primary** Energy Per Capita 10^{10} Joules	Primary Energy Savings Per Capita 10^{10} Joules	1975 Total Primary Energy 10^{10} Joules	2050 Total Primary Energy 10^{10} Joules
Residential	6.9	2.4	4.5	14.6	6.7
Commercial	5.9	1.8	4.1	12.6	4.9
Transportation	9.2	2.3	6.9	19.6	6.5
Personal	(6.0)	(1.4)	(4.6)	(12.7)	(3.8)
Freight	(3.2)	(1.0)	(2.3)	(6.9)	(2.7)
Industrial***	13.0	6.2	6.8	27.6	17
TOTAL	35.0	12	22	74.4	35

2050 primary energy per capita is 36% of 1975 primary energy per capita.
2050 total primary energy is 47% of 1975 total primary energy.

SOURCE: 1975 total primary energy from News Release March 14, 1977. "Annual U.S. Energy Use Up in 1976," Office of Assistant Director—Fuels, Bureau of Mines, U.S. Department of the Interior.

NOTES: The division of energy into residential and commercial from Bureau of Mines figures follows Amory B. Lovins. "Scale, Centralization and Electrification in Energy Systems," a paper presented at Oak Ridge National Laboratories, October 1976.

Totals may not add due to rounding.

*213×10⁶ population.

**277×10⁶ population.

***Miscellaneous and unaccounted for in Bureau of Mines data included in industrial.

chapters 4 through 9, including the winding down of the automobile, road-building, and food-processing industries; and the rise of the solar energy, mass transit, telecommunications, housing, construction, and health care industries. In chapters 10 and 11, we discuss shifts in employment patterns and in the distribution of work, leisure, and income. These trends anticipate the end of the continued growth of the commodity component of society's output. The final sections treat the policy questions, issues, and measures, as well as the international implications, that are consistent with this scenario.

I. The Personal Sector

3. *Settlement Patterns and Changing Communities*
4. *Residential Energy Use*
5. *Food and Agriculture*

3. Settlement Patterns and Changing Communities

MANY PROBLEMS of the man-made environment are problems of size. Getting out of a city of 100,000 people is relatively easy, even without an automobile; to get beyond the edge of a city of three million, even with an auto, takes a major effort. Air pollution in a city would be less of a problem if it were dispersed over a large area or if cars were restricted in numbers or number of miles travelled—the seriousness of the problem is a function of city size. The magnitude of the problem of solid-waste disposal more than doubles with a doubling in the size of a city, not because the volume of waste more than doubles, but because all of it must be hauled farther for disposal. Economies can result from small scale as well as large.

Today's pattern of extremely high and extremely low densities would, in our scenario, evolve into a pattern of medium-size cities with moderate densities, separated not by suburban sprawl but by farmland and forest. Such cities would be small enough to avoid megalopolitan diseconomies of scale yet large enough to support a vigorous industrial and commercial base and sustain a high level of cultural activity. Charles E. Little (1975, p. 8) has described such a future urban pattern:

> Instead of the seemingly endless sprawls we now know, suburbs should be villages, composed of a series of hamlets, closely spaced, that together make up a village and support the services a village can provide. . . . The village may, within its political boundaries, contain, say, eight square miles, but only three square miles should be developed. The average density of the developed area should be five dwelling units per acre. The village would have a population, then, of about 35,000. Five square miles should be field and forest, mountain and river. . . . The five square miles should, of course, link up with the open spaces of neighboring villages in a linear fashion that leaves

ondal Partridge

undisturbed and undeveloped the major landscape features, including the principal drainage ways, the ridgelines of hills, and outstanding scenic areas. The village should own all this land. Much of it should be put to agricultural or silvicultural use on a lease basis. . . . Many of the prospects, the striking visual landscapes, may once have been sites for those 1950s-style subdivisions of half-acre, single-family houses. They should be razed.

Impossible to achieve?, asks Little, after conjuring up this image. Probably so, given the style of development made possible by the private automobile. Probably not, if development is required to accommodate constrained private transportation.

City planner Patrick Geddes (1915) thought that decentralization would be brought about by the development of electricity that would free industry from centralized power facilities. Ralph Borsodi theorized in the 1930s that it could be achieved by the proliferation of power tools that would make possible home-based skilled labor. His theory was based on the sound but economically heretical assumption that it is more economical to haul machine parts than workmen. Frank Lloyd Wright felt that the expansion of the automobile would bring it about and designed Broadacre City on that premise. But none of the technological innovations that could have been the basis for social innovation halted the trend toward centralization and urbanization. The social, cultural, and economic attraction of the city has overwhelmed any competing considerations. Put simply, modern industrial culture has been rich enough to afford the kind of cities it has, even though there have been simpler and more forthright means of providing for human needs.

The costs of Barbara Ward's "unintended city" (1976), not planned for human purposes but shaped by the hammers of technology, applied power, the overwhelming drive of self-interest and the single-minded pursuit of economic gain, are now glaringly apparent in the form of urban crime, bankruptcy, blight, and environmental deterioration. Catering to the automobile, modern commerce has substituted the air-conditioned shopping mall surrounded by black-top for the shop-bordered town square or piazza as a place for people to meet and mingle. This new town center is now far afield and stripped of all community functions but the sale of commodities. The energy, environmental, and financial costs of these changes suggest that a metamorphosis to new settlement patterns has become necessary.

In our scenario, these new forms would be characterized by smaller cities of more human scale with present megalopoli broken up into communities of 50,000 to 100,000 inhabitants—such as England's "urban villages" of Chelsea and Trastevere or New York's Greenwich Village (Ward, 1976). We foresee a density along the lines suggested by Little (1975), roughly five dwelling units per acre. The population distribution we envision reflects the form of our proposed future settlements (Table 2).

The overall effect on America's large cities would be the simultaneous implosion and explosion of population. The implosion would occur as the automobile was abandoned as the primary means of city transit. The explosion would come about with the reestablishment of distinct—as opposed to megalopolitan—urban areas of reduced size. These areas would generally contain 200,000 or fewer residents, a population still manageable in terms of the economics of city operation, full pedestrian and bicycle access to all districts, and accessibility to surrounding food-producing lands (Bell, 1973). A fundamental aim in redesigning urban areas would be maximum accessibility as opposed to maximum mobility. As private transportation becomes more and more costly, and as local grocers, butchers,

TABLE 2. U.S. DEMOGRAPHY 1970 AND 2050

	1970				2050			
	Number of Cities	Average Population	Total Population	Percent of Total	Number of Cities	Average Population	Total Population	Percent of Total
Total U.S. Population			203.2				277	
Cities>100,000	156	362,000	56.5	28	100	200,000	20	7
Cities>25,000–100,000	760	45,500	34.6	17	1,500	53,000	79	29
Cities 2,500–25,000	5,519	7,682	42.4	21	8,000	10,000	84	30
Rural including urban <2,500 and incorporated urban			69.8	34			94	34

SOURCE: Bureau of the Census, Department of Commerce, Statistical Abstract of the United States, 1977, U.S. Census, Series II-X projection 1980–2050.

bakers, druggists, jewelers, booksellers, restauranters, and clothiers are able to supply most of local residents' needs, the neighborhood and the town would regain their social and community functions. It may be that the parking lots of today's shopping centers will be the sites of tomorrow's housing.

Evolving from the present to this future state has eluded planners, government officials, and citizenry who have attempted to face the problems of present settlement patterns. We believe, however, that the transformation is possible; our belief is supported by recent data on the depopulation of the largest urban areas. Migration out of these areas is already occurring.

The history of the United States is sometimes written in terms of population migration; there is no evidence that these migrations have concluded. Figure 1 displays the familiar westward and southward migrations. The westward movement has persisted from the nation's earliest days. Southward migration, recently popular as a theme for social and political analysis, has been steady since 1910 and has accelerated substantially since 1940. Superimposed on these changes has been migration from rural areas to cities, from small towns to large metropolitan areas, and, within the metropolitan areas, the movement of residences from city centers to suburban areas. But between 1970 and 1974 an unanticipated movement began. Ten of the thirteen largest megalopolitan areas (the Standard Consolidated Statistical Areas of the U.S. Census Bureau) exhibited net population migration out of the urban areas. Figure 2 shows the net migration as a function of city size.

The changes implied by these trends are not predicated on comprehensive planning but on the way many people have chosen to cope with the realities of high-cost energy. The shifts we foresee in settlement patterns (Table 2) would result from the encouragement of these trends.

Legislative action would be a part of that encouragement, facilitating the abandonment of nonviable populated areas, primarily in the megalopoli, as well as the construction of both rural settlements and viable communities in small cities. Although we do not foresee the construction of many complete new towns, we do anticipate the incremental rebuilding of established towns.

A major step in repopulating rural America would be a National Homestead Lease Act, permitting individuals to lease, at no cost,

FIGURE 1. POPULATION MIGRATION WEST AND SOUTH

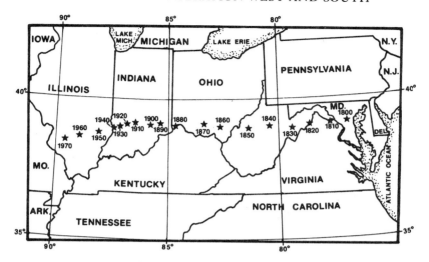

SOURCE: U.S. Bureau of Census, Department of Commerce.
U.S. Bureau of Population, 1970. Vol. 1.

FIGURE 2. NET POPULATION MIGRATION FOR THE LARGEST MEGALOPOLI
(STANDARD CONSOLIDATED STATISTICAL AREAS)

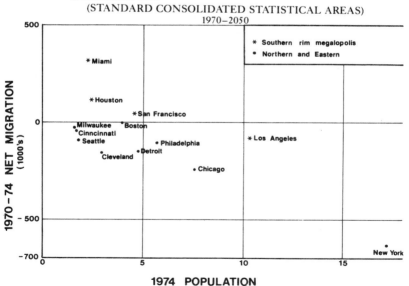

SOURCE: U.S. Bureau of Census, Department of Commerce.
Statistical Abstract of the United States, 1977.

tracts of publicly owned land, provided the lessee makes his home on the land. Tracts might be four to forty acres in size, depending upon the location. Leases might be renewable indefinitely, as long as the lessee continued to live on the tract and cultivated some portion of it. Incentives and assistance might be offered at the outset. We estimate that one to four million people would take advantage of such a program, which would cost the federal government less than it now spends on job programs. In fact, if two million took advantage of the program, all their land needs could be met by public land in unused and abandoned military bases.

An indication that the federal government has given careful thought to the depopulation of rural areas and the general destruction of rural community life that has occurred in the past half-century is the Family Farm Act of 1974. Designed to restrict the corporate industrialization of American agriculture, the Act prohibits any corporation with nonfarm assets of more than three million dollars from owning or operating farms (Goldschmidt, 1975; U. S. Senate special Small Business Committee, 1946). Enacted at a time when American agriculture is on the verge of being captured by corporate interests, this law provides needed protection for the farmer-owner—with no loss of efficiency. The U.S. Department of Agriculture, usually the ally of large-scale farming, concluded in a 1967 study that the most efficient organization in American agriculture is the one- or two-family farm (Madden, 1967).

A repopulation of rural America would take some pressure off the largest cities and preserve the good aspects of city culture that have been destroyed by urban growth far beyond the human scale. U. S. Census data for recent years indicate that the four fastest growth rates in the United States are in rural areas. Substantial migration out of urban areas, which we have included as part of this scenario, would leave those who stayed behind in possession of cities that could be made very livable.

As a counterpart to the National Homestead Lease Act, Congress could expand the National Urban Homestead Act with the three-fold purpose of:

- providing housing for those in the cities who cannot fully afford it;
- improving the inner cities by putting to use structurally sound buildings now falling into disrepair whose replacement would require impossible amounts of both capital and energy;

• doing these things in a way that would increase the independence and equity of the individuals involved.

The guiding principle of the law is the idea elaborated nearly a decade ago by George Sternlieb (1966): ownership is the basic variable accounting for differences in the maintenance of slum properties. Sternlieb's idea for renewal, tried successfully on a small scale by a number of cities in the 1970s, consists of selling publicly owned, abandoned dwellings to families for a low price in return for the new owner's agreement to bring the building up to city codes within 18 months and to occupy it for at least five years. Because few of the prospective owners would be able to obtain loans to finance the rehabilitation of their buildings, public low-interest loans and loan guarantees would be made available on the basis of maintaining future property values and the income arising from the improvements made. The abandonment of inner-city buildings has been a growing urban problem of some time; in 1975, some 30,000 buildings were abandoned in New York City alone. Many of these buildings are actually structurally sound but badly in need of repair. They would be ideally suited for this renewal plan and also could be made to serve the goal of energy conservation. For example, in 1976, the Community Services Administration funded an experimental program in which heavy insulation and a solar water-heating system were installed in an abandoned five-story, eleven-unit tenement as part of the "sweat equity" renovation of the building by its new owners.

As a necessary complement to the rural repopulation and urban rebuilding programs, a National Conurbation Act could provide for the restructuring of the present megalopolis. The Act's intent would be to develop a regional pattern of distinct cities and towns that would be designed as no-growth entities and would include farms and open space. As people move out, abandoned unsound structures would be demolished with nothing built in their place. In large cities, buildings, and in some cases whole blocks, have been left vacant. New York City, for example, now has more than one square mile of vacant office space (3 km^2). In slum neighborhoods out of necessity, in better-off neighborhoods out of frugality or civic pride, these vacant areas could be converted into community gardens and orchards. The cities, straining under the loss of tax base as a result of

]26[

A family farm, agriculture's most efficient organizational unit. Photo by Vermont Agency of Development and Community Affairs.

out-migration, would welcome this use of empty lots. After a time, the trend of building rehabilitation, urban open-space accrual, and new building construction would lead to multi-nucleated patterns of compact towns and communities on the sites of present-day megalopoli. Such planned development, with appropriate zoning ordinances, would integrate the energy, agricultural, transport, and recreational needs of a community while avoiding unnecessary transportation of food and materials from distant regions. Zoning in certain areas would allow a mix of commercial and residential uses on the same block, and in the same building complex, so residents would need to travel only short distances to shop and to their place of work.

Thus, we envision recent trends, with encouragement and structuring from the legislation, leading to a restructuring of settlements.

4. Residential Energy Use

S PACE AND WATER heating consume about seventy-five percent of the total United States energy used in the residential sector (USDI, 1975). The rest is used for cooling,
lighting, air conditioning, and a variety of appliances.

As many citizens have noticed in recent years, significant savings
in heating fuels can be obtained by changing the allocation of heat to
a home both spatially and temperally (zoned control, day and night
thermostat setbacks). An estimated twenty-five percent, and often
more, of heating fuel costs can be saved (Pilati, 1975). We expect this
trend of changing behavior during the heating season to continue,
since fuel prices most likely will continue to rise.

Major reductions in both heating and cooling requirements can be
obtained by retrofitting old buildings and carefully designing new
buildings. Reducing infiltration, adding insulation, and improving
windows of old buildings could result in an average reduction of at
least fifty percent in heating requirements. (Note: percentage reductions by behavioral and structural changes are not additive.)
Renovation of old buildings is consistent with a policy of moderate
urban density without autos, since most of the older structures are
relatively closely spaced. Up to eighty percent reduction in heating
needs is possible for new buildings by building smaller homes and by
using sound building techniques and correct window placement to
take advantage of solar energy (AIA, 1975; Bliss, 1976).

Water-heating devices are presently very inefficient and can take
up to fifteen percent of total residential energy demand. Improved
insulation, lower temperature setting, heat recovery from other appliances, and reduced use of hot water can lower this energy consumption by fifty percent or more (Ross and Williams, 1976).

We estimate that the implementation of these measures would
decrease the energy consumption per capita for space and water
heating to about 30 to 40 percent of the 1975 level by the year 2050

and in many cases would reduce it to an even smaller percentage. Total residential energy use per capita can, therefore, be reduced by about 50 to 60 percent by the year 2050. It is worth noting that space and water heating require relatively low-temperature heating devices and can, therefore, be met in large part by simple solar (and wind) heating systems.

FIGURE 3. SPACE AND WATER HEATING–COMMERCIAL AND RESIDENTIAL

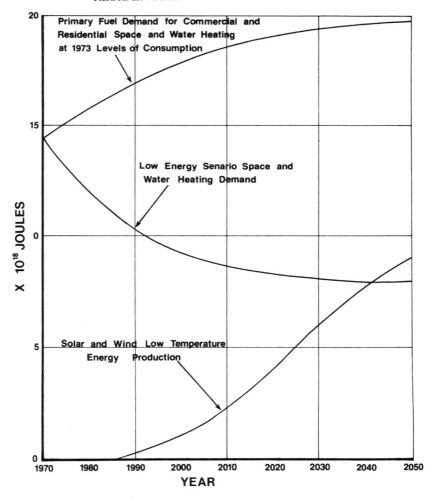

About twenty-five percent of residential energy use is for cooling, refrigeration, and air conditioning (Steinhart and Steinhart, 1974). Increasing the efficiency of appliances would have a significant impact. Furthermore, better design of new buildings would make air conditioning unnecessary in a large part of this country by the year 2050, although energy use for home freezers would increase somewhat because of the changing food system. We estimate that energy use for all of these devices could be cut by about fifty percent, or about a ten percent reduction in the total residential energy use. Electricity would continue to be the major form of energy used for these tasks, with the possible exception of air conditioning.

Energy-efficient building codes on the national, state, and local levels, and national and state efficiency standards and labeling laws for energy-using equipment would help to bring about these reductions in residential energy use.

The combination of the measures mentioned above would lower the residential energy use per capita from about 6.9×10^{10} joules in 1975 to about 2.4×10^{10} joules in 2050—a reduction of 65 percent. Total residential energy use for the year 2050 is estimated at approximately 6.7×10^{18} joules, of which about 75 percent, or 5.0×10^{18} joules would be for space and water heating (low-temperature energy needs).

Figure 3 displays the demand for space and water heating for *both* the residential and commercial sectors between 1975 and 2050.

5. Food and Agriculture

THIS SCENARIO foresees a changeover to regional agriculture as well as changes in the average American diet that would reduce total energy use in food production by fifty percent over the next seventy-five years.

Until the 1940s, most regions of the United States produced most of their own food, specialized large-scale food production being an invention of the last twenty-five years. The startling growth in home gardening and local direct-marketing cooperatives in the early 1970s suggest that a trend away from centralized and specialized food production may have already begun. Large-scale crop production for export will be with us as long as developing nations struggle with food production, but with rising energy prices the United States will be able to remain competitive only by decreasing the energy subsidy to these crops. And food in the United States is already expensive by world food-cost standards.

The present fossil-fuel subsidy to the American food system in the form of gasoline, machinery, fertilizer, and pesticides is fifteen calories for every calorie of food energy consumed (Booz-Allen, 1976). As fossil fuels become scarce, agriculture of this intensity will become impossible, not only in the developed nations upon whose agriculture much of the world has come to depend, but also in the less-developed countries whose hopes for progress rest on boosting food production with machinery and chemical fertilizers.

The American diet and food system, according to our scenario, would have to change over the next seventy-five years. Beef consumption would be cut by half and the beef that is eaten would be produced from pasture- and range-fed cattle eating grasses that humans cannot digest instead of from feedlot cattle raised on valuable grain. Swine would be fed poor quality grain and refuse. Land formerly used to grow feed for both cattle and swine would be used to grow grain and beans for human consumption, much of which would be exported. As beef protein intake diminished, consumption of other foods would increase—foods such as fish, chicken, eggs, soybeans, and dairy products, all of which have high protein-conversion efficiency and lower energy requirements.

Fish would be supplied by coastal and local fishing and the extensive use of fish farms. Poultry and egg production would operate on a local basis in a far more energy-efficient manner than current poultry factories. The consumption of dry beans and fresh potatoes, which has declined in the United States in recent decades, would increase, as would the direct consumption of grains. The consumption of fruits and vegetables—ninety percent of which would be grown and eaten locally—would rise.

Even if direct consumption of grain were to increase to make up for the reduced meat intake, and even if per-acre grain production were to drop slightly as a result of the less energy-intensive use of chemical fertilizers, there would still remain available a substantially increased quantity of grain for export (Chapman, 1973). A reduction in American meat consumption, however, would not be designed to provide more grain to feed the world's hungry, although such grain surpluses would provide reserves for year-to-year fluctuations in production as well as for famine-stricken areas. Even if enough grain could be freed this way to supply the world, it would be just one more stopgap measure, since population growth would eventually exhaust these grains as well. For Americans, the dietary changes would reflect market responses to escalating prices of energy-intensive foods. Maintaining moderate food costs is of considerable interest nationally, even for upper-income families who could pay higher prices, for spending on non-food items would have to decline as food prices rise.

In addition to anticipating an American diet that would include a greater proportion of whole grains, dried beans, potatoes, and vegetables generally, we expect that most produce would be grown for local consumption by individuals, by local gardening cooperatives, or by local farmers. It is not difficult to imagine a time when lawns would have very nearly disappeared and suburban houses would be surrounded by gardens and orchards, their residents having gradually come to refuse to pay for high-priced commercial produce. Areas that now look like green residential deserts would resemble the intensely farmed and carefully tended gardens of England or China.

Young worker in an inner city community garden, San Francisco. Photo by Rose Skytta, Jeroboam, Inc.

The most fascinating thing about a widespread resurgence of gardening would be the huge amount of food produced by what seems to us today so much like an avocation. Truly, no one should be surprised by the economic value of home gardens; the U.S. Department of Agriculture has calculated that a family of four could be well fed without animal products by using only one-sixth of an acre.

Modern intensive gardening techniques produce still higher produce-yields per unit area.

Both the National Conurbation Act and the National Homestead Lease Act would support the trends to more local food production and distribution. Energy savings from these measures, and in the agricultural sector (see Steinhart and Steinhart, 1974), contribute to the reductions in industrial, residential, commercial, and transportation energy use summarized in the overview table. The National Container Act would bring about further savings in the food system.

II. The Business Sector

6. *Commercial Energy Use*
7. *Transportation*
8. *Industry*
9. *Energy Supply*

6. Commercial Energy Use

THE CHARACTERISTICS of energy consumption in the commercial sector are similar to those of the residential sector. Space and water heating represent about seventy-five percent and air conditioning about ten percent of total commercial energy use, all of which need only low-temperature energy sources. Conservation measures and legislation outlined for the residential sector would reduce commercial energy consumption dramatically. In addition, significant savings would be made by using "waste" heat generated within the buildings. Smaller, nondetached stores and other moderately sized commercial buildings would make extensive use of natural ventilation. These changes would result in a seventy percent reduction in commercial energy needs for space heating and cooling and water heating, or about a sixty percent reduction in total commercial energy use per capita by the year 2050.

Additional energy savings would occur through the use of natural light and reduced lighting standards. A reduction in energy use for lighting would increase heating needs, but these needs could generally be met without the use of electricity. Reduced use of lighting for advertising would have the same effect. Furthermore, changes envisioned in the food system would have a significant impact on commercial energy use: phase-out of fast-food services and other "junk" commerce, a drastic reduction in packaging through a shift to standard sizes, and more careful display of goods (e.g., no open freezer cases). The popular fact that the McDonald's hamburger chain uses enough energy in a year—largely for packaging—to provide electricity for the cities of Pittsburgh, Boston, Washington, and San Francisco combined, illustrates the American penchant for throwaway containers (Ingersoll, 1972). As energy becomes more costly, a society that uses packaging and produces garbage on the scale of the United States will find the reduction of this waste to be a necessity.

Energy use for packaging would decline as disposable paper and plastic containers nearly disappear, as commodities that could be shipped, stored, and merchandised in bulk are handled that way, and as nonessential junk foods, overprocessed foods, and excessive toiletries and cosmetics nearly vanish. The National Container Act

Photo by Kent Reno, Jeroboam, Inc.

would reduce both waste and the energy used for packaging. Litter and solid waste would cease to be a problem when commercial packaging and containers are designed for nearly complete recycling and reuse. Moreover, since about one-eighth of local trucking is for garbage and trash disposal, reduced trash would mean less trucking.

We estimate that the combination of these measures would reduce commercial energy use per capita from 5.9 x 10^{10} joules in 1975 to 1.8 x 10^{10} joules in 2050—a reduction of approximately 70 percent. Total commercial energy use for the year 2050 is estimated at about 4.9 x 10^{18} joules, of which 70 percent, or 3.5 x 10^{18} joules, is for space and water heating (low-temperature energy needs).

7. Transportation

TRANSPORTATION energy savings in personal travel and freight hauling described in this scenario would reduce direct energy use per capita to twenty-five percent of present levels. What is more surprising is that no reduction in social interaction or accessibility would occur. The energy savings would result primarily from two factors. First, because redesigned settlements would reduce the length of most car trips and provide alternative modes of transportation, including safe walking and bicycle pathways, the necessity of driving would be markedly reduced. (Work trips have accounted for the largest share of trips, approximately 35 percent; business trips for 15 percent; school, 8 percent; social and recreational, 20 percent; and other, 13 percent [U.S. Department of Transportation, 1974].) Secondly, for the remaining trips that require driving, the vehicles involved would be smaller and far more efficient in their use of energy per vehicle mile. The same factors would apply to energy used for hauling freight.

The change foreseen in settlement patterns would not reduce the number of trips people would make, but rather reduce the distances spanned and change the modes of transportation. Assuming, for example, a city of circular form with five dwelling units per acre, or 10,000 people per square mile, the distance from the edge of the city to the center would be one mile for a city of 31,000—a twenty-minute walk or a six- to ten-minute bicycle ride. The distance would be two miles for a city of 126,000. This would reverse the trends in social and physical arrangements that began with the widespread ownership of the automobile and the universal willingness to sacrifice 25 to 50 percent of urban areas to the operation and storage of the automobile. The pattern of placing numerous competing food and department stores together in shopping centers and malls accessible only to private autos would be replaced by the neighborhood store. Schools and work places would also be distributed throughout the communi-

Photo by Rondal Partridge

ty. With the emphasis on walking, bicycle, and mass transit, far less land would be given over to roadways, parking, and the associated noise and emissions. The net effect would be to allot the five dwelling units per acre even more space than under present conditions.

These factors, partially motivated by the increasing cost of operating automobiles, would reduce average automobile mileage per capita from 8,000 miles to 3,000 miles. After accounting for increased use of other motorized forms of transportation, we anticipate a fifty percent reduction in energy use from these measures.

Precedents for this type of transportation future already exist and others are being developed. In Rotterdam, 43 percent of all daily trips are by bicycle. The cities of Runcorn, England (population 70,000) and Port Grimaud, France, are both planned to function without automobiles, providing instead exclusive bus, bicycle, and pedestrian access, except for emergencies and special deliveries.

To eliminate the least efficient aspects of air travel, this scenario envisions a heavy emphasis on rapid intercity rail systems, especially for trips under five hundred miles. Buses and the automobile, how-

]41[

ever, would continue to carry a significant proportion of intercity travelers. The requisite interlinking of these modes, especially at terminals, would be part of the National Transportation Coordination Act. Other acts in the scenario would call for the termination of the Interstate Highway Program (Interstate Highway Termination Act), the termination of air terminal expansion and other airline subsidies (Airline Deregulation Act), and the rebuilding and subsequent expansion of the existing rail network (Railroad Revitalization Act). This would be a long-term process and would involve some nationalization to provide the necessary capital.

This scenario assumes that energy-efficient speed limits would be enforced for auto, air, and truck travel. We also anticipate the passage of a Fuel Economy Act that would call for slightly more stringent standards than those required by the Energy Policy and Conservation Act of 1975, which mandates 27.5 miles per gallon for the 1985 new car fleet average. (In recent years, this average has been around 11 to 15 miles per gallon.) Under this act, cars would average 2,200 pounds instead of the current 4,000 pounds, would obtain an average of at least 25 mpg in town, and would last 20 years or 200,000 miles.

The fuel used for transportation would shift away from an almost exclusive dependence on petroleum to a mix of electricity, synthetic fuels, hydrogen, and petroleum. Twenty-five percent of energy used for personal travel would be electric, mostly for local automobile trips and electrified rail and bus travel. Fifty percent of the energy used for hauling freight would be electrified, with electric trains on all main rail lines and electric trucks for local delivery.

These changes in vehicle characteristics would result in a reduction of over fifty percent in the energy used per vehicle mile. The combined effects of reduced vehicle mileage and improved efficiency would result in a seventy-seven percent reduction in the energy used annually for personal travel—from 6.0×10^{10} joules per capita in 1975 to 1.4×10^{10} joules per capita in 2050.

Similar savings would be realized in the hauling of freight. Although recreation areas and farmland would separate communities, ton-miles would drop dramatically because of the decentralization and localization of food production and processing (at present about one-half of all trucks haul food and agricultural products). More decentralized manufacture in industry and the shifts in production noted in the industrial sector, combined with the changes in the agricultural sector, would result in a fifty percent reduction in ton-miles. [One ton of freight moved one mile = 1 ton-mile.—Ed.]

Citywide delivery systems would be put into operation as part of the modal shift to the use of railroads and the decline of centralized shopping centers. Intelligently done, the distribution of goods to neighborhood stores should involve little more time or cost than delivery to today's centralized shopping centers. This is especially true given a shift away from trucking freight cross-country directly to the central retailer from the producers, a practice not found in rail-supplied cities where local distribution from the city railhead to neighborhood outlets is done by small trucks.

In addition, we estimate that with ninety percent of all air freight eliminated and seventy-five percent of all truck freight shifted to rail carriage, the total amount of energy used in freight transportation would be reduced forty percent.

The combined effects of reduced ton mileage, modal shifts, and increased modal efficiences, especially for trucks, would result in a seventy percent per capita reduction in the energy used to move freight—from 3.2×10^{10} joules per year in 1975 to 1.0×10^{10} in 2050. As an illustration of how such a reduction could be achieved, the Cummins Engine Company of Columbus, Indiana, a major manu-facturer of truck engines, has estimated that a twenty-five percent savings is possible with a typical tractor semitrailer combination. With an appropriate engine, transmission, and axle combination, thirty-five percent could be saved.

Total transportation energy use, including both freight and per-sonal, would be 2.4×10^{10} joules per capita in the year 2050.

8. Industry

INDUSTRIAL production currently uses thirty-seven percent of the total energy budget of the United States, or 27.6 x 10^{18} joules annually (Ross and Williams, 1976). This scenario envisions changes in the economy's mix of goods and services that would reduce production in certain energy-intensive industries (Tables 3 and 4) for a saving of 3.8 x 10^{10} joules per capita—or 10.9 percent of total 1975 energy use. (This estimate does not include savings from resulting cutbacks in the energy-supply industry.) Overall output of goods, though changing in mix, would cease to grow on a per capita basis.

Table 3 summarizes information contained in *A Time to Choose*, the report of the Energy Policy Project of the Ford Foundation (FFEPP, 1974, pp. 140–157), as an aid in determining the effect of changing the output mix and associated production cutbacks in certain energy-intensive sectors. An example of such a cutback would be a decline in automobile production due to decreased auto ownership and use and greater auto durability. To derive the effects of such changes, each of the energy-intensive industries in Table 3 is hypothetically cut back thirty percent, with the exception of food and kindred products and the primary metals industries, which are reduced fifty percent. Changes in the food and kindred products industries would primarily occur in processing. The change in primary metals would be largely due to changes in automobile production.

These changes, of course, would be due to production changes, not to conservation. Conservation measures are separate measures that can stand by themselves or be undertaken in addition to the hypothetical cutbacks. Table 4 also assumes that overall levels of activity in the rest of the economy would remain as they were in Table 3. This assumption cannot be defended except as an heuristic device for interpreting the magnitude of the effects. By the same token, the cutbacks do not take into account secondary effects. With these

TABLE 3. ECONOMIC PARAMETERS OF ENERGY-INTENSIVE INDUSTRIES, 1971

Energy-Intensive Sectors	% of Manufacturing Gross Energy Consumption	% U.S. Employment	% Industrial Production	% New Plant and Equipment Investment
Primary metal	26.8	1.6	6.3	3.4
Chemical and allied	17.0	1.1	9.4	4.2
Stone, clay, and glass	7.3	.8	2.9	1.1
Paper and allied products	7.6	.9	3.6	1.5
Food and kindred products	6.3	2.1	9.6	3.3
Subtotal	67.8	7.3	31.8	13.6

TABLE 4. ECONOMIC PARAMETERS OF TABLE 3 INDUSTRIES, AFTER PRODUCTION CHANGE

Energy-Intensive Sectors	% of Manufacturing Gross Energy Consumption	% U.S. Employment	% Industrial Production	% New Plant and Equipment Investment
Primary metal	19.3	.8	3.7	1.8
Chemical and allied	16.7	.8	7.7	3.1
Stone, clay, and glass	7.2	.6	2.3	.9
Paper and allied products	7.5	.6	2.9	1.2
Food and kindred products	4.5	1.1	5.6	1.7
Subtotal	55.2	3.8	22.2	8.7

TABLE 5. PROJECTED ENERGY SAVINGS BY CONSERVATION MEASURES IN THE INDUSTRIAL SECTOR

MEASURES	% REDUCTION
Good housekeeping measures through industry (except for feedstocks)	13
Fuel instead of electric heat in direct heat applications	.6
Process steam and electric cogeneration	8.7
Heat recuperators or regenerators in 50% of direct heat applications	2.5
Electricity from bottoming cycles in 50% of direct heat applications	1.7
Recycling of iron and steel in urban refuse	.4
Reduced throughout at oil refineries	3.0
Reduced field and transport losses associated with reduced use of natural gas	2.7
TOTAL REDUCTION is about	33

SOURCE: Estimates are based on data from Ross and Williams (1976).

NOTE: Assumption is made that the percentage reductions would be approximately the same after projected changes in industrial production.

qualifications, the cutbacks noted will:

1) save 29 percent of manufacturing gross energy consumption, or approximately 9.7 percent of total energy use;
2) add 3.4 percent to unemployment (This would be more than made up for by increased employment in new energy sources and to carry out conservation measures [U.S. Congress, 1978]. (See below.)
3) reduce industrial production 12 percent;
4) save or make available 5.5 percent of investment capital for new plants and equipment.

Aside from these changes in industrial production, a significant reduction in industrial energy consumption would be obtained by efficiency improvements. Strenuous energy conservation measures subsequent to production changes would reduce energy use by 3.0 x 10^{10} joules per capita, an additional thirty-three percent. We estimated this future energy saving by using data from Ross and Williams (1976), who based their study largely on the concept of the second law of thermodynamic efficiency as outlined by the American Physical Society (1975). Higher efficiencies would be obtained by carefully matching the energy quality—or temperature of the energy supply—to the demand. The introduction of relatively small-scale multipurpose power plants, as we suggest in the chapter on energy supply, would play a major role in this context. We expect that fuel-use taxes would be placed on industries in the near future to discourage the use of oil and natural gas (Depletable Fuels Tax Act), making multipurpose power plants economically attractive. Table 5 gives our estimates of how conservation measures would reduce energy consumption in the industrial sector by the year 2050.

The total reduction in industrial energy use resulting from changes in production and the adoption of energy conservation measures would be 6.8 x 10^{10} joules per capita, a fifty-two percent drop from the 1973–1975 level of industrial energy use. By 2050, total industrial primary energy use would be 17 x 10^{18} joules.

9. Energy Supply

DECREASING per capita energy consumption by sixty-four percent would postpone but not solve our energy supply problems. Increasing population, combined with drastically decreasing production of oil and natural gas, will put considerable pressure on other energy supplies. Thus, in this scenario, solar and wind sources would supply a large fraction of our energy needs. Alternative supplies such as wood, bio-fuels, geothermal, etc., would also grow substantially. Coal, with its associated environmental problems, would, in 2050 and beyond, be used in amounts smaller than at present.

The 2050 electric generating system would be substantially different from that of today. We also envision the decline of the nuclear industry. By the mid-1980s, rising costs and public concern over safety would halt nuclear power plant construction, including construction of breeder reactors, already plagued by technological problems, concern over the threat of nuclear proliferation, and problems of cost. Solar, wind, hydro, and geothermal electric generating systems would supply the majority of the electric demand. Marginal-cost pricing, combined with other load management techniques, would be used to tailor the demand curve to follow supply. Daylight peaks might be encouraged to make use of the cheapest solar production station. As cryogenic storage and electrolysis become less expensive, solar may assume a base-load function. Hybrid plants such as coal-solar or coal-wind would be built while wind would be used to generate electricity on a centralized as well as on a decentralized basis. The trend in coal plants would be toward small (less than three hundred megawatts) multipurpose plants. Industries would generate their own process heat and electricity, selling the excess as district heating expands. The transition to the proposed system would be gradual, new plants of the type described replacing currently operating plants as they become obsolete. Diverse and innovative electric generation would be the result.

Construction of a solar house in a solar subdivision, Davis, California. Photo by
William Rosenthal, Jeroboam, Inc.

Several factors would facilitate this transition. As we've noted
above, we expect to see nuclear construction moratoria, an end to
nuclear subsidies, and an end to the federal commercial breeder
program. Some of these policy changes are already receiving strong
public impetus through local opposition to nuclear plants. The cost of
solar and wind energy systems, both thermal and electric, would drop
as a result of improved techniques and mass production. In some
cases, demonstration projects would provide the necessary production demand. A tax and loan-guarantee incentive program for the use
of renewable (flow) resources on state and federal levels would help
this transition. A national strip-mining act would set stringent standards for land restoration so that the continued use of coal would not
seriously scar the landscape. Finally, a progressive depletable-fuels
tax would be levied on most uses of oil and gas as fuels, rising steeply
for those uses for which there are alternatives. The naturally increasing costs of depletable fuels, combined with this tax, would
improve the economic viability of alternative energy sources.
viability of alternative energy sources.

In the future described in this scenario, multipurpose power stations would generally be used in industrial and moderate- and high-density residential and commercial areas. At the present time, the industrial sector consumes about one-third of the electricity generated in the United States. This electricity is mainly used for motors, lighting, and electrolysis, which are high-quality energy needs. In addition, about forty-five percent of all industrial energy is expanded to generate process steam. Electricity could be produced by the industries in combination with steam generation. Gyftopoulos et al. (1974) calculated that with the present requirements for process steam enough electricity could be produced as a by-product not only to meet the electricity needs of industry, but to have a surplus to sell. Similar power plants could supply both electricity and heat to residential and commercial sectors, where they can be adapted to specific energy needs of the area. A general increase in fuel prices would, we expect, make these small but relatively efficient "total energy systems" cost advantageous in the near future. We estimate that coal would be used to fire many of these plants in the

n in New Hampshire, a home using a combination of collector panels, green-se, and southern glass wall, consumes one-fifth the energy of a standard house. to by Helene Kassler.

industrial sector and some, as well, in the residential and commercial sector. Total coal consumption would be slightly lower than total industrial energy consumption by the year 2050.

Solar heating for space and water will, we think, be in widespread use by 2050 and will supply, together with wind, most of our low-temperature energy needs. The rapid growth in solar heating has already begun. In the last two years, the number of firms engaged in solar energy production has grown by a factor of ten. Figure 3 illustrates our requirements for space and water heating and our estimates for low-temperature solar and wind applications. By the year 2020, nearly all buildings will have solar or wind heating for space and water. Moreover, approximately sixteen percent of total United States energy use is for industrial process heat below 350 degrees Farenheit (*Machine Design*, 1976, p.4), which is easily accommodated by solar and wind technologies.

We foresee that thermal electric generation from central solar stations will be available on a commercial scale by 1985. Early solar plants will most likely be coupled with coal-fired plants, but these hybrid plants will eventually give way to solar base-load production when cryogenic storage and electrolysis become economical.

One of the few technological breakthroughs we see in our future is the declining cost of photovoltaic cells. We forecast that by 1990 the

TABLE 6. PHOTOVOLTAIC COST PROJECTIONS

YEAR	COST PER PEAK KILOWATT (1976 DOLLARS)	REFERENCES
1959	$200,000	Chalmers 1976
1976 (March)	21,000	Machine Design 1976
1976 (October)	15,500	Machine Design 1976
1979	5,000	Machine Design 1976
1986	500	Chalmers 1976
		Machine Design 1976
		Electronics 1976

FIGURE 4. TOTAL SOLAR ENERGY SUPPLY PROJECTIONS:
1970–2050

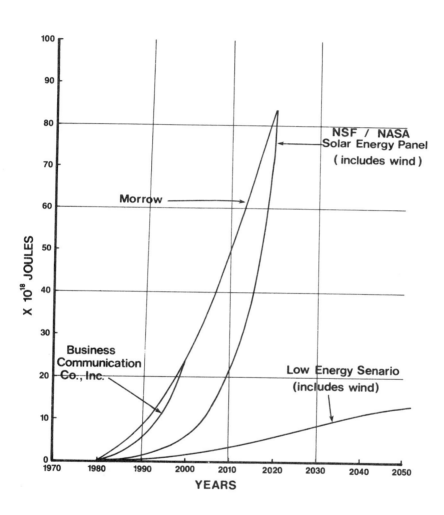

SOURCES: Business Communications Co., Inc. 1975. "Solar Energy: A Realistic
Source of Power." Stamford, Connecticut.

Morrow, W. E., Jr. 1975. "Solar Energy: Its Time Is Near" in *Perspectives on Energy*,
Ed. Ruediselli, Lon C. and Firebaugh, Morris W. New York: Oxford University Press,
pp. 336–351.

NSF/NASA Solar Energy Panel. 1971. "Solar Energy as a National Energy Source."
College Park, Maryland.

cost of photovoltaic electric generation will be $1,000 (in 1976 dollars) per installed kilowatt of capacity—a little less than the current cost for nuclear electric plants. We base our estimate on the current trend in photovoltaic costs, the expected breakthrough in the EFG (edge-defined film growth) production technique of silicon cells, and the refinement of gallium arsenide (GaAs) cells that are capable of high efficiencies (twenty percent) in high-temperature applications (*Computer Design*, 1975). Table 6 shows current and projected photovoltaic costs for panels only. Figure 4 is a graph of total solar energy supply projections.

Wind energy in central as well as dispersed applications could supply a significant part of the electrical capacity in 2050. Included in such wind systems would be generators ranging in size from a few kilowatts for individual homes to installations of one to two megawatts connected into electrical grid systems. Capital costs per installed kilowatt capacity of wind machines and conventional electric generators are presently of the same order of magnitude. Wind electric systems with relatively short-term storage systems or an array of wind machines spread over a large geographical area appear to have approximately the same reliability as large conventional generating units. (Sorenson, 1976). Wind energy conversion systems could also serve as space heating devices for a large geographical area of the United States, since the seasonal availability of the wind corresponds with space heating needs.

Present trends of rapid growth in electric space heating could continue, to some extent, with large wind electric systems, if such space heating were combined with centrally controlled heat storage devices in individual buildings. (Fifty percent of new dwellings in 1976 have electric heating devices. In certain areas of the Midwest, a fifteen percent annual increase in the number of all-electric homes in not uncommon.) In addition, individual wind machines for neighborhoods and small towns could be used as "wind furnaces," without any interconnection to the electrical grid. Combinations of these wind-thermal and electric systems could also be employed. The economics of wind furnaces appear to be as good as solar heating devices and no major technical breakthroughs are needed for large-scale applications of wind energy conversion systems. Table 7 illustrates the role of wind energy conversion systems as a major energy source in the future.

TABLE 7. OVERVIEW OF PROJECTED ENERGY SUPPLY, 1975–2050 (10^{18} JOULES)

	1975	1980	1990	2000	2010	2020	2030	2040	2050
Solar electric, including photovoltaic	–	–	.10	.20	.43	.80	1.1	1.5	1.9
Wind electric	–	–	–	.44	.80	1.1	1.5	1.8	2.0
Hydro and geothermal	3.3	3.4	3.5	3.8	4.1	4.4	4.7	4.9	5.0
Solar and wind heating	–	–	.25	1.0	2.3	4.0	6.0	7.6	9.0
Coal	14.1	15	19	26	27	25	21	17	13
Oil—imported*	12.9	12	7.0	2.0	–	–	–	–	–
Oil—domestic	21.4	21	14	11	9.5	7.0	5.5	3.4	2.0
Synthetic liquids**	–	–	–	.10	.12	.20	.50	1.2	2.0
Natural gas	21.1	21	15	8.3	4.0	1.8	.80	.44	.25
Nuclear electric generation***	1.6	1.6	.95	.37	.15	–	–	–	–
Total	74.4	74	60	53	48	44	41	38	35

NOTE: Figures may not add up because of rounding. For estimated future amounts only two significant figures are presented.

*No oil for U.S. imports available approximately 10 years after expected world peak production.

**Synthetic liquids from coal, oil shale, bioconversion, and others will increasingly be used for transportation, combined with domestic oil.

***Last nuclear power plant construction in 1985; approximate lifetime 30 years.

Table 7 and Figure 5 summarize energy supply in various stages in the development of our scenario. We have separated low- and high-temperature energy supplies to further illustrate the necessity of thermodynamically matching supply and demand.

FIGURE 5.

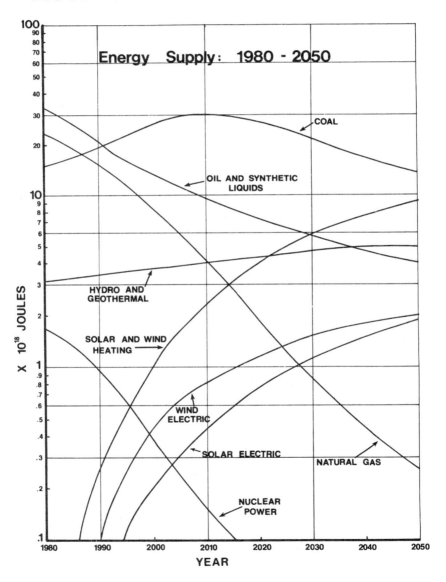

III. Government, Politics, and the Public Sector

10. Employment, Social Welfare, and Leisure Time

A NY ATTEMPT to control energy consumption by restraining economic output will naturally enough raise cries of opposition from those concerned with maintaining employment, increasing profits, and getting a piece of the economic pie—labor, industry, and the poor, respectively. Yet Sweden provides an example of how stable employment can be maintained during a time of reduced energy supplies and subsequent economic reorganization. By "rationalizing" its industry—i.e., carrying out structural changes in industries and adapting new techniques of production—Sweden has been able to maintain a highly efficient economy. This has been possible, according to Gunnar Myrdal (1972), because a firm national policy of full employment in the form of guaranteed jobs, effective retraining programs, and smooth relocation of labor has to all intents and purposes banished the threat of technological unemployment. Were the United States to take full employment seriously and guarantee jobs to everyone wanting them, organized labor's fears of economic reorganization could be eased and the transition to a low-energy economy made smoother.

In this scenario, we have set the average workweek in 1990 at thirty hours to reflect two changes: a twenty-five percent decrease in work time for the average individual; and a twenty-five percent decrease in real income that is offset by an increase in potentially profitable leisure time. A thirty-hour workweek would then be available for a large bloc of the structurally unemployed, who would fill the work hours left by reduced work time. Reducing average work hours to seventy-five percent of present levels would presumably provide employment opportunities for one new worker for every three now employed, assuming that there was no change in total production and labor productivity. The 1976 labor force of 84 million could thus be expanded by some 28 million, just to maintain current output. A Full Employment Act would provide the needed impetus

Fixing up older buildings and retrofitting them with solar equipment would generate thousands of jobs. Photo by Peeter Vilms, Jeroboam, Inc.

by establishing a thirty-hour workweek for federal and state jobs by 1985.

How, it will be asked, can we be so sure that people will agree to lose one-quarter of their income in exchange for more free time? We can't be sure, of course, but we may at least interpret, with Dennis Gabor (1972, pp. 10–11), the increased and endemic employee absenteeism of recent years as a sign that a significant bloc of people have already reached the point where time off means more to them than the dollars they forego in not going to work. The opportunity to work a shorter week has never before existed except for low-paying part-time jobs: one is usually expected to work the customary full week or not at all.

The increased time available may indeed prove more profitable than the equivalent time spent on the job, for some people would use the unpaid time off not just for leisure, but to perform services for themselves. Conceivably, this would result in the service being completed more cheaply than if the person remained on the job during that period and paid to have the service done.

Would the commercial and industrial establishment be able to handle the increased numbers of employees? We believe so, since the 1976 autoworker's contract effectively provides nearly one-fourth of the year as paid time off.

Institutionalizing the option to take one's wealth in the form of leisure time and individually performed service time instead of in goods may be the way for a society to smooth the transition to a period of reduced energy availability. Finally, a society that is at last forced to substitute the serious redistribution of wealth for economic growth as a means of eliminating poverty may find that spreading aggregate labor hours among individuals by reducing the average workweek is a palatable means of accomplishing that end.

Substantial interregional and interindustrial shifts of labor would take place as part of the adjustment to a low-energy society. The transfer of population and industry from the North and Northeast to the South and Southwest will continue (chapter 3), bringing the new industrial states into greater prominence. Factors that encouraged this shift—favorable tax structures, lower wage rates, and a generally unorganized labor force (Koeppel, 1976)—will produce a less than acceptable situation for workers until union strength is sufficiently developed and state regulatory authority effectively applied. The Appropriate Skills Retraining Act would ease this transition by providing increased federal support for trade and technical schools that teach needed new skills. The demand for new housing could be treated as an opportunity to install well-built, energy-efficient homes for the influx of workers. At the same time, this drain of people from the older metropolitan areas would open up an opportunity to renovate and rehabilitate older cities, increasing the demand for the labor of those still living there.

Interindustry shifts of labor in our scenario would be significant. The housing industry, actively trying to fill the need for energy-efficient dwellings to replace wasteful older buildings, would grow still further in response to the need for reconstruction, rehabilitation, insulation, and solar retrofitting. A report prepared for the New

York Legislative Commission on Energy Systems noted that conservation measures would lead to increased employment. By saving one million kW (1000 MW) through energy conservation at $300 per kW, the employment level would rise by 12,000 people on a thirty-year cycle. This is slightly less than a nuclear or coal plant would employ to generate 1,000 MW, but significant numbers of people would be employed immediately in the conservation plan. (Included in these figures is the labor required for maintenance, material processing, component manufacturing, and construction, but not mining.)

tretch of the suburban Boston mass transit system. Photo by Rondal Partridge

Efforts to conserve energy in existing buildings would employ electricians; carpenters; plasterers; painters; truck drivers; factory workers; engineers; heating, ventilating, and air-conditioning specialists; and other building-trade workers. As investment per conserved kilowatt increases, so does employment. Conservation of energy would not, it seems, reduce employment, as many have claimed, but would instead increase it (Monroe, 1976).

The rehabilitation of railroads and the construction of mass transit systems would also require an increased labor force. Bezdek and Hannon (1974) report the following: total employment would increase and energy use would be reduced if the highway trust fund was reinvested in any of several alternative federal programs. If railroad and mass transit construction monies were shifted from highways to railroads, the energy required for construction would be reduced by about sixty-two percent and employment would increase by 3.2 percent. Passenger transport by railroad required more labor and less money and energy than car transport in 1963.

A similar conclusion was reached in an Energy Research Group study (1973) that substituted buses for automobiles in urban areas. Freight transport by railroad was less expensive in dollars, energy, and labor requirements than was truck transportation in 1973. If the monetary savings had been absorbed as a tax and respent on railroad and mass transit construction, about 1.2 million jobs would have been created. Had there been a full shift from intercity car and truck transportation to transportation by railroad, with savings spent on railroad construction, 2.4 million new jobs could have been created in 1963. This is the trend we envision for railroad and mass transit use and construction.

The manufacture and installation of centralized as well as single-unit solar and wind energy systems would, in the scenario's future, be a burgeoning industry, successfully taking up the slack labor from declining automotive and fossil fuel industries. The exception would be coal, which would grow in the interim to supply that part of electrical demand not met by renewable resources. Wind generators require no fuel, but the operation and maintenance of a large wind system would require, on a continuous basis, two to four times the labor force as would an equivalent nuclear or coal plant (Monroe, 1976). Increasing reliance upon solar energy is likely to have substantial employment benefits. Even though some of the components would undoubtedly require larger production facilities, many of the collectors could be produced locally, and certainly installed locally, with relatively small capital investment. The installation of solar energy systems is a labor-intensive process. The Federal Energy Administration's Project Independence Task Force estimates that the solar energy industry could create half a million jobs by the end of this decade (U.S. House Committee on Science and Astronautics, 1974).

The need for labor in recycling would increase as this industry expanded to handle not only metals and glass but also wooden, paper, and cloth containers and to transfer urban sewage sludge to agricultural and silvicultural uses. This new employment would take up the slack created in the throwaway packaging and container industry.

As energy becomes more and more costly, labor would return to agriculture, horticulture, and silviculture, though in new, less physically demanding roles. We noted earlier some of the changes that may take place in an economy working under the new constraints of energy supply and we suspect that the makeup of organized labor would be altered as well.

11. Policy Implications and Implementation

HOW ARE THE ECONOMIC and social changes we forecast to come about? Existing trends and logic will not be enough. Public and private policy changes are going to be necessary. Although government intervention in our lives has come to be resented, we cannot escape the fact that in the complex society in which we find ourselves, governmental meddling may be a necessary evil. The specific measures that we have proposed are summarized in Table 8. Many of them have already been considered in some form, and although none require massive political changes, a good number will not see easy passage. No attempt has been made to be all-inclusive in describing policies that would have an impact on energy use. For example, defense programs can be very energy intensive but have not been included.

Direct as well as indirect impacts of these policies must be considered, for their effect on our social, economic, and political environment is crucial. For example, how is the striving for material gain to be relaxed? Will people acquiesce quietly to high prices and reduced material consumption? Why will individuals cease to be acquisitive? We do not suppose that they will, but we suspect that they will gradually accommodate their wants to diminished means. Economic wants are not unlimited, and we do not agree with the premise of the industrial revolution that all wants expand indefinitely. We would like to think that the "drive beyond consumption" noted by W. W. Rostow is becoming stronger and that the pursuit of Maslow's "higher needs" is the message of the 1960s (Maslow, 1964, 1971). All of this would ease the transition we envision, but in the end we rely on a conventional mix of economic and legislative measures to accomplish, or at least give impetus to, many of our proposed changes.

Economic growth has been prescribed as a means of redistributing income and eliminating poverty through the provision of jobs. Robert Lampmann (1972) argued in the late 1960s that economic develop-

TABLE 8. POLICY DIRECTIONS AND TIMETABLE

POLICY AREA	DESCRIPTION	IMPLEMENTATION LEVEL	TIMETABLE
POLICY AREA	DESCRIPTION	IMPLEMENTATION LEVEL	TIMETABLE
Transportation	Railroad Revitalization Act—regulatory reform and partial nationalization	National	1978–2000
	Fuel Economy Act—higher fleet requirements for miles/gallon. Tax penalties for poor economy, and strict speed enforcement	National and state	1976–1985
	Interstate Highway Termination Act—end of federal highway building program; redirecting of funds to mass transit and railroads	National	1978–1985
	National Transportation Coordination Act	National	1978–1985
	Airline Deregulation Act	National	1978–1979
Energy and Fuel Supply	Nuclear power moratoria	National, state, and local	1977–1985
	End of federal nuclear subsidies	National	1977–1985
	Solar and wind incentive programs—tax writeoffs, loans, research, product standards, education, and demonstration projects	National and state	1977–1990
	National Stripmining and Restoration Act	National	1977–1980
	Depletable Fuels Tax Act	National	1977–1990
	Synthetic Fuels Standards Act—quality and transportation standards	National	1985–1990

Category	Policy	Scope	Years
Economic Reform	GNP redefinition incorporating externalities	National	1977–1985
	Guaranteed Income Act	National	1980–1985
	Corporate Reform Act—effective antitrust regulations, more progressive income tax, incentives for small businesses	National and state	1978–1985
	Stabilization of world prices through indexing of resource prices to prices for industrial products and food	International	1980–1990
Energy and Resource Conservation	Energy efficient building codes	National, state, and local	1977–1980
	Efficiency standards and labeling laws—includes all energy using equipment	National and state	1977–1985
	Marginal cost pricing for electricity	National and state	1977–1985
	National Container Act—standard, recyclable containers; state versions also	National and state	1977–1985
Industrial	Multipurpose generation laws to allow the cogeneration and sale of steam, heat, and electricity by industries	National and state	1980–1990
	Utilities cooperation acts to facilitate public ownership and energy sales to utilities	National and state	1977–1995
Employment	Full Employment Act—30-hour workweek for federal and state jobs	National and state	1980–1985
	Appropriate Skills Retraining Act—increased federal support for trade and technical schools teaching needed new skills	National	1977–1985
	Appropriate Technology Act—funding for appropriate technology research and development	National	1977–1985
Land Use	National Homestead Leasing Act	National	1980–1985
	National Urban Homestead Act	National	1980–1985
	National Conurbation Act	National	1990–1995
	Mixed use zoning and taxation laws	State and local	1980–1990

ment had improved the general lot of the poor. But for a significant segment of the population, a good part of the time, the usual means of securing an income is not available: there are no jobs. Even in boom times there stubbornly persists a class of "structurally unemployed" living on some form of relief and usually in slums. Adherence to the idea of aggregate economic growth may have its justifications, but the belief that it helps to alleviate poverty is among the poorest. The "trickle-down" theory, if it works at all, is neither fast enough nor equitable enough to meet the needs of all people in the short time we have available to make the transition to a low-energy economy. In a future of energy and materials austerity, the sort of GNP growth we as a nation have come to expect will be impossible to maintain, and will hardly lend itself to the redistribution of income.

We are more optimistic than Lord Keynes, who noted with dread "the readjustments of the habits and instincts of the ordinary man, bred into him for countless generations, which he may be asked to discard within a few decades." We do not think a calamitous transition is inevitable so long as no individuals or segments of society are grievously displaced or left destitute and so long as the populace generally feels that all are bearing their share of the complications, inconveniences, and troubles of the conversion to a low-energy economy. Destitution alone has never been a cause for revolt; history seems to indicate instead that poverty combined with rising expectations is the formula for revolution. By the same token, a middle class fearful of losing jobs and income and persuaded by demagogues that both domestic and foreign conspiracies are causing and profiting from their hardship is the recipe for an ugly strain of fascism. Individual economic security must be assured. Once it has been, the gradual erosion of the commodity standard of living may be accommodated in the same way that fiscal inflationary erosion is accommodated now.

The value of a guaranteed annual income (GAI) in a world of limited growth may be greater than any of its early proponents could have suspected. Supporters of the GAI continually explained how it would increase the opportunities for study, social services, and environmental improvements (Johnson, 1971). One can also imagine its prospective value as a damper on economic growth. An income program that adequately supplied what we now refer to as necessities, combined with the option of additional work to pay for luxuries, might generate a class of people who would be willing to

The "trickle-down" theory, an unassailable belief for some, has done little for society's marginally employed. Photo by George Ballis.

live at a lower income level and forego these luxuries if to get them meant more work. The obvious outcome of this is reduced economic demand and reduced energy demand.

If we are able to assure everyone's economic security, what then? We may take some direction from Abraham Maslow's theory of a hierarchy of needs (1964, 1971). He believed, as does Marianne Frankenhauser (1974), that as physical needs are met, higher needs such as love, learning, and self-actualization become our driving forces. We have difficulty talking about transcending needs in the United States, a nation at the pinnacle of material wealth, although its cities rot, thousands of its citizens cannot find decent housing, and many adults and children suffer from malnutrition and lack of basic medical care. Our policies must take the inequities of our present society into account, since only when we are all assured of the basics of life may our higher needs be met. In this process our traditional beliefs, economics, and politics will undoubtedly have to change.

12. International Considerations

THE MOST OBVIOUS international consequence of this scenario is the reduction in a gradual and orderly fashion of United States payments for foreign oil imports. Only the increase of arms sales abroad from about one billion dollars in 1970 to more than ten billion dollars in 1976 has prevented far more serious balance-of-payment deficits. How long can we, or should we, be the world's leading arms supplier? The end of United States oil imports will come shortly after 2000 (Table 7), about the time world oil production is expected to peak and begin a decline (Hubbert, 1975, p. 113).

Of course, the national security arguments for ending oil imports would be served by this scenario, but considering the dependence of the United States on foreign sources for more than thirty critical raw materials, we have never been persuaded by these arguments.

The nonnuclear future of this scenario offers several international advantages:

· The effort begun by President Carter to eliminate proliferation of nuclear weapons via nuclear power would be possible without agreements or inspections;
· The United States, in trying to persuade others to forego nuclear enrichment technology, would be believable by example;
· Nuclear waste disposal and nuclear accidents would disappear as problems;
· The United States could become the technological leader in solar and wind technologies.

Increasing world oil prices of the early 1970s were an even greater burden to the poor nations of the world than to the United States. With limited foreign exchange and development plans resting on oil-dependent technology, many poor nations have accumulated growing debts that may threaten the international banking system.

Using wind and solar energy and food grown on the premises, the Ark Project on Canada's Prince Edward Island is a prototype for integrated energy-food systems. Photo by Stewart Brand.

At best, development plans may be delayed by the drain on foreign exchange generated by the need for oil. Are we really helping such nations by urging them to adopt energy-intensive agriculture and by offering them nuclear power plants and other costly and energy-intensive technologies?

Opinions vary widely about the responsibility of the industrialized world to assist third world development, but few would defend policies that inhibit attempts by poor nations to accomplish their own development. If, as poor nations strive to better their lot, we are perceived as outbidding them for scarce resources, conflict is a likely result. How much better to offer developed solar, wind, and geothermal technologies that may be adapted more easily to labor-rich third world countries than to offer capital-intensive nuclear plants or oil-dependent machinery. In any case, eliminating United States fuel imports would help preserve remaining stocks for developing nations.

We have no illusions about any voluntary redistribution of the world's wealth, but this scenario offers the chance for a future profitable to both rich and poor. The alternative may mean confrontation between the rich few and the destitute many.

Further Considerations

Epilogue

References

About Friends of the Earth

Epilogue: Three Years Later

Three years ago we began serious efforts to assemble a comprehensive low-energy scenario. During a graduate seminar on energy policy in the fall of 1975, conversation kept returning to energy policy and its purposes. The conventional forecasts of energy needs were extrapolations of the past with minor modifications, suggesting the future was to be like the past—only more so—a most unlikely outcome if the present was any index. Cities needed to change past patterns and improve their decaying cores; minorities and the poor hoped and worked for change; 30 years of farm programs had been drastically reduced—portending changes; a national commitment was under way to reverse a century-long decline in air and water quality, despite continuing conflict about how rapidly and vigorously to pursue these programs; and, seemingly quite apart from programs and policies, demographic changes were underway that were accompanied by obvious changes in lifestyle. In short, whether we looked at the turbulent present, at political platforms, or at what ought to be, what we saw were changes that—whatever else might happen —were not extrapolations of the past.

More specifically, domestic oil and gas production had peaked and seemed hardly likely to go anywhere in the future except down. Project Independence as policy was blandly oblivious to much of the politics of the real world. (If, after all, we achieved energy independence, what were our foreign policy partners in Europe and Asia to do?) As a last gasp of extrapolation policy, Project Independence was headed for a fast trip to the science-fiction shelf. The scenarios of the Ford Energy Policy Project had hardly been published before the highly optimistic estimates of recoverable oil and gas in the United States that underlay them were revised drastically downward by the U.S. Geological Survey, which had provided the estimates in the first place.

It had long been clear that, as a purely economic matter, conservation of energy was the cheapest incremental way of coping with energy shortages. It was also the only thing that could be done quickly. What was the focus of vigorous dispute in 1975 was how much conservation and how fast. The American Physical Society summer study that year had presented the possibility of dramatically reducing energy use by a better matching of energy expenditure to the specific tasks to be performed. The Energy Policy and Conservation Act of 1975 was under debate and ultimately passed the Congress, instituting a schedule of automobile fuel efficiency, among other provisions. The estimates of energy conservation, the lengthy reports on energy policy, and even the laws passed, all dealt in part, and implicitly, with changes in the way we live; on the more explicit question of how things ought to be, however, they were mostly silent. The noisy public discussion was concerned more with placing blame for problems than with visions of how to solve the problems.

In this setting, we began, with our fellow students in the seminar, to generate a picture of how little energy use might be contemplated without either basic changes in human nature or astonishing technological breakthroughs. The result in the earlier pages is just one of great many scenarios that could have been generated with these two strictures. We found we could not avoid a number of subjective choices for which science provides no guidance. In discussion we began to understand that a "value free" scenario was quite impossible, and that forecasts that purport to be value free simply choose to hide the assumptions at an earlier and unstated stage of the work. We take this opportunity to thank our critics as well as our sympathizers in the seminar for their contributions, especially in requiring us to face up to the difficult problems of transition to any future more sane than the present.

All the major trends we spoke of appear to be continuing: settlement patterns are changing, with emigration from the big cities of the North and Northeast and the continued growth of the Southern Rim cities; rural towns are growing (some unhappily, as resource exploitation operations overwhelm them), and renovation is reviving the pedestrian city. Minneapolis is one example of a city with a declining population whose remaining residents continue to rehabilitate old houses and grow gardens on vacant lots. Other anecdotal examples come from Boston, San Francisco, Philadelphia, and Washington, D.C. But in some other areas exurban development continues wildly and major projects for rehabilitation of urban centers are not always successful. There are gradually spreading signs of concern over uncontrolled sprawl and at the same time the basic extensions of roads, sewers, water and other services have been subjected to more scrutiny and citizen input almost everywhere.

One of the strong points of our scenario (we think) is that we tried to look at solving energy problems by looking at the secondary and tertiary influences on energy use: don't just make cars more efficient, obviate, as well, the need to use them as much by building residences near workplaces, relocating, when necessary, commercial and recreation areas. The axiom we adopted, "maximum accessibility, not maximum mobility," remains an imperative. We spoke of the transformation and decline of the giant malls that have sprung up around the country and that are generally accessible only to autos. Since we wrote about that problem, the shopping center and "mall" have come

orting bottle returns in Oregon, where, by law, all bottles must be returnable. Photo by Peeter Vilms, Jeroboam, Inc.

under increasing attack, as formerly bustling downtown areas (primarily of smaller cities) are drained of consumers and hard times beset the remaining businesses.

The Homestead Lease Act and other measures to accommodate "back to the land" folks seem far from enactment and some examples of such efforts appear discouraging despite occasional successes. But population flow into small towns and rural areas continues—in North Carolina, New England, Wisconsin, Arkansas, Colorado and elsewhere.

Perhaps we will see the development of villages and towns (with all the modern telecommunications and other community services and amenities) that both house the farmers and provide agricultural logistical support, such as cooperative machinery operation and maintenance. The farmers would fan out each day to work their individual acreage, fully backed by the resources of the Organic Agricultural Extension Service Agricultural Cooperative. Passage of the Cooperative Banking Act in 1978 has made this or other innovative local solutions more likely.

We still insist that neighborhood stores are fine and necessary in the pedestrian city, but much more is needed to broaden consumer access to a wide range of goods and services. We note, for instance, that doctor's house calls are making a comeback; and phone-order grocery service is again available in most large cities. (Minneapolis has one that is competitive with the supermarkets in both quality and price; and of course they deliver.) The advent of the moderate-cost home computer could enable the people in our scenario to enjoy the consumer accessibility they have today without most of the commuter inconvenience they now experience. Two-way cable T.V. is a reality now in Ohio, with the QUBE system, and in England, with View-Data, and seems finally ready to penetrate the mass markets of the world.

The automobile seems to be with us for a long time to come, but the fleet fuel standards should produce significant reductions in gasoline use at least by the mid-1980s. Meanwhile residential and commercial electricity conservation has exceeded utility company expectations, , resulting in delay or cancellation of a number of scheduled power plants. As a result of this and other causes, nuclear power plant orders have dropped dramatically worldwide since 1975, reinforcing our belief that a scenario for a non-nuclear future is realistic.

]77[

by Stewart Brand.

A low-energy future for the United States is still a possibility. Nationally, the marked decline in the growth of electricity demand and the forecasts for future demand have been paralleled by reductions in growth of other fuel use. Conservation is evident in each of the demand sectors. America is only now beginning to realize its conservation potential. Awareness of cost savings and fuel savings is being spurred by both petroleum price increases and electric rate increases. Even from the last three years it has become increasingly evident to us that the growth pathways of the past have been permanently abandoned.

Of the legislative timetable, 5 of the 28 suggested measures have been or are being acted upon: stripmining protection, energy efficient building codes, appliance efficiency standards, appropriate technology, and airline deregulation (though still retaining vast subsidies to airport facilities at several government levels). A national container act is even under consideration. The fate of the Crude Oil Equalization Tax of the Carter energy program (roughly equivalent to our depleteable fuels tax act) suggests that many legislative measures may have a far rockier road.

Among our second thoughts is a growing feeling that we made too much of legislating the future. As many have done, we may have outlined too large a role for the government. Grass-roots activity is appearing everywhere, though statistics are hard to come by.

Another common error we share with statisticians and economists is the excessive reliance on averages, large numbers, and statistics generally. The future, like the present, holds diversity and attendant conflicts over objectives and lifestyles. The real challenge to the government may well be to live with this diversity and conflict—and to derive political choices from it—so that we have a chance to foster any ideas that will work. Our social fabric will be similarly challenged as people with divergent views of the future try to work and live together.

Finally, in the area of energy supplies, advances in alternative sources seem to be outstripping government goals. Even Energy Secretary Schlesinger, a long-time nuclear advocate, has increased his estimates of the solar contribution during the next 20 years. Nevertheless, one of the most sobering parts of preparing an energy scenario is designing a transition from a society based on oil and gas to one in which oil and gas play a minor role. Vigorous debates over electric generation continue, but the problems lie in the need for

The early versions of the scenario were presented first to an energy workshop in Europe and then to a meeting of the American Association for the Advancement of Science. Scientists, especially those in the energy area, responded with polite indifference. By contrast, journalists snapped up the available 200 copies and many more asked where to write for additional copies. Since then, demand for the scenario has been steady from government agencies, planning groups, citizen's organizations of various kinds, and from abroad. In the last year or two there has been a growing interest (but not necessarily agreement) from scientists as well. The scenario has been translated into three foreign languages that we know of.

The widest exposure of the scenario has come through several feature articles in newspapers and two local television specials in the Midwest. In that sense, we feel that we have been very lucky, for our aim was to bring others into the arena of public discussion of possible futures. These earlier notices as well as this publication by Friends of the Earth provide just such an opportunity. If there is a message in all this it is to hold on to your dreams. Expert advice aplenty will be needed, but the shape of our future is a legitimate concern for us all.

J.S., M.E.H., C.C.deW.,
R.W.G., K.B.L., M.T.,
S.J.K.

Madison, Wisconsin
January, 1979

REFERENCES

American Institute of Architects. 1975. *A Nation of Energy Efficient Buildings by 1990*. Washington, D.C.

American Physical Society. 1975. *Efficient Use of Energy*. A physics perspective: a report of the summer study on technical aspects of efficient energy utilization. In ERDA Authorization-Part 1, 1976 and Transition Period Conservation. Hearings before the Subcommittee on Energy Research Development and Demonstration of the Committee on Science and Technology. U.S. House of Representatives. Ninety-fourth Congress, 1st session, February 18.

Bell, Daniel. 1973. *The Coming of Post Industrial Society: A Venture in Social Forecasting*. New York: Basic Books.

Bezdek, Roger, and Hannon, Bruce. 1974. Energy, Manpower and the Highway Trust Fund. *Science*, Vol. 185, 23 August. pp. 669–675.

Bliss, Raymond W. 1976. Why not just build the house right in the first place. *Bulletin of Atomic Scientists*, Vol. 32, No. 3, March. pp. 32–40.

Booz-Allen Associates. 1976. *Energy Consumption in the Food System*. Report No. 13392-007-001. Washington D.C., USGPO.

Business Communications Co., Inc. 1975. *Solar Energy: A Realistic Source of Power*. Stamford, Connecticut.

Chalmers, Bruce. 1976. The Photovoltaic Generation of Electricity. *Scientific American*, October. p. 34.

Chapman, Dwayne. 1973. An End to Chemical Farming. *Environment*, March. pp. 12–18.

Computer Design. September 1975. Gallium Arsenide Use for Low-Cost High-Efficiency Solar Cells. Concord, Mass. p. 44.
Electronics. April 1, 1976. Highstown, New Jersey.

Energy Research Group. 1973. *Urban Auto-Bus Substitution: The Dollar, Energy and Employment Impacts*. Report to Energy Policy Project. 1776 Massachusetts Avenue, NW, Washington, D.C., in press. As referenced in Bezdek and Hannon (1974).

Ford Foundation Energy Policy Project (FFEPP) 1974. *A Time to Choose: The Nation's Energy Future*. Cambridge, Massachusetts: Ballinger Publishing Company.

Forrester, Jay W. 1971. *World Dynamics*. Cambridge, Massachusetts: Wright-Allen Press.

Frankenhauser, Marianne. 1974. Limits of Tolerance and Quality of Life. Paper presented at the symposium, Level of Living-Quality of Life, held at Biskops—Arno, Sweden. Published in *Viewpoint*, the Swedish Information Service, New York, September.

Gabor, Dennis. 1972. *The Mature Society*. London: Secker and Warburg.

Geddes, Patrick. 1915. *Cities in Evolution: An Introduction to the Town Planning Movement and to the Study of Civics*. London, England: Williams and Norgate.

Goldschmidt, Walter. 1975. A Tale of Two Towns. In *Food for People, Not for Profit*, ed. by Lerza and Jacobson. New York: Random House.

Gyftopoulos, Elias P. et al. 1974. *Potential Fuel Effectiveness in Industry*. A report to the Ford Foundation Energy Policy Project. Cambridge, Mass.: Ballinger Publishing Company.

Hayes, Dennis. 1976. *The Case for Conservation*. Washington, D.C.: Worldwatch Institute.

Herrera, Amilcar et al. 1976. *Catastrophe or New Society: A Latin American World Model*. Ottawa, Canada: International Development Research Centre.

Heilbronner, Robert. 1974. *An Inquiry into the Human Prospect*. New York: W. W. Norton and Co.

Hubbert, M. K. 1975. Survey of World Energy Resources. In *Perspectives on Energy*, ed. by L. C. Ruedisili and M. W. Firebaugh. New York: Oxford University Press.

Ingersoll, Bruce. 1972. *Chicago Sun Times*, October 30. Article reports in some detail Professor Bruce Hannon's study of the McDonald's hamburger chain.

Johnson, Warren A. 1971. The Guaranteed Income as an Environmental Measure. In *Economic Growth vs. the Environment*, ed. by Warren A. Johnson and John Hardesty. Belmont, California: Wadsworth Publishing Co.

Koeppel, Barbara. 1976. Something Could Be Finer than to be in Caroline. In *The Progressive*, June. pp. 20–23.

Lampman, Robert J. 1972. As cited in W. W. Heller, Coming to Terms with Growth and the Environment. In *Energy, Economic Growth and the Environment*, ed. by S. H. Schurr. Baltimore: Johns Hopkins University Press.

Little, Charles E. 1975. The Double Standard of Open Space. In *Environmental Quality and Social Justice in Urban America*. Washington, D.C.: The Conservation Foundation.

Lovins, Amory B. 1976. Energy Strategy: The Road Not Taken? *Foreign Affairs*, Vol. 55, No. 1 pp. 65–96.

Machine Design. October 7, 1976. Penton Publishing Co., Inc. Cleveland, Ohio. pp. 4, 10.

Madden, Patrick J. 1967. Economies of Size in Farming: Theory, Analytical Procedure, and a Review of Selected Studies. USDA Economic Research Service. Agricultural Economics Report 107. Washington, D. C.

Maslow, Abraham. 1964. *Religions, Values, and Peak Experiences*. Columbus, Ohio: Ohio State University Press.

Maslow, Abraham. 1971. *The Further Reaches of Human Nature*. New York: Viking Press.

Meadows, Donella H.; Meadows, Dennis L.; Randers, Jorgen; Behrens, William W. III. *The Limits to Growth: A Report for the Club of Rome's Project on the Predicament of Mankind*. New York: Universe Books.

Mesarovich, Mihajlo, and Pestel, Eduard. 1974. *Mankind at the Turning Point: The Second Report to the Club of Rome*. New York: E. P. Dutton and Co., Inc./Reader's Digest Press.

Monroe, James et al. 1976. Energy and Employment in New York State, Draft Report. A Report to the New York State Legislative Commission on Energy Systems. Report ES 119. May 3.

Morrow, W. E., Jr. 1975. Solar Energy: Its Time is Near. In *Perspectives on Energy*, edited by Lon C. Ruedisili and Morris W. Firebaugh. Oxford University Press, New York. pp. 336–351.

NSF/NASA Solar Energy Panel. 1972. Solar Energy as a National Energy Source. College Park, Maryland.

Pilati, D. A. 1975. *The Energy Conservation Potential of Winter Thermostat Reductions and Night Setback*. Oak Ridge, Tennessee: Oak Ridge National Lab.

Ross, Marc H., and Williams, Robert H. 1976. Energy Efficiency: Our Most Underrated Energy Resource. *Bulletin of the Atomic Scientists*, November. pp. 30–38.

Sorenson, Bent. 1976. *On the Fluctuating Power Generation of Large Wind Energy Converters, with and without Storage.* Copenhagen, Denmark: Niels Bohr Institute, University of Copenhagen.

Stavrianos, L. S. 1976. *The Promise of the Coming Dark Age.* San Francisco: W. H. Freeman and Co.

Steinhart, S. E., and Steinhart, J. S. 1974. *Energy: Sources, Use and Role in Human Affairs.* North Scituate, Massachusetts: Duxbury Press.

Sternlieb, George. 1966. Slum Housing: A Functional Analysis. *Law and Contemporary Problems.*

Vacca, Roberto. 1974. *The Coming Dark Age.* Garden City, New York: Anchor Press.

Ward, Barbara. 1976. *The Home of Man.* New York: W. W. Norton and Co.

Watt, Kenneth E. 1974. *The Titanic Effect.* Stamford, Conn.: Sinauer Associates, Inc.

U.S. Bureau of the Census. Department of Commerce. U.S. Census of Population, 1970. Vol. 1; *Statistical Abstract of the United States.* 1977.

U.S. Congress. 1978. *Creating Jobs through Energy Policy.* Hearings before the Subcommittee on Energy of the Joint Economic Committee. Ninety-fifth Congress. March 15 and 16.

U.S. Department of the Interior (USDI). 1975. Energy Perspectives; A Presentation of Major Energy and Energy-Related Data. Washington, D.C.: Government Printing Office.

U.S. Department of Transportation/Federal Highway Administration. 1974. Report 10. *National Personal Transportation Survey: Purposes of Automobile Trips and Travel.*

U.S. House Committee on Science and Astronautics. 1974. Subcommittee on Energy Hearings, 30 July and 2 August, on Solar Energy Research, Development, and Demonstration Act of 1974. pp. 170–363. (Report of Subpanel IX of FEA's Project Independence Task Force. Solar Energy and Other Sources. Prepared for Chairman, Atomic Energy Commission. October 27, 1973.)

U.S. Senate Special Committee to Study Problems of American Small Business. 1946. 79th Congress. Second Session. Senate Committee Print. No. 13. Report Pursuant to Senate Resolution 28. December 23, 1946. Small Business and Community; Study in Central Valley of California on Effects of Scale on Farm Operations.

THE AUTHORS

John S. Steinhart received degrees in economics and geophysics and is professor of geology and geophysics and of environmental studies at the University of Wisconsin-Madison.

Mark E. Hanson has degrees in economics, water resources, and land resources. He is currently holder of a Rockefeller Foundation post-doctoral fellowship in environmental affairs and is associated with the Energy Systems and Policy Research Group at the University of Wisconsin-Madison.

Carel C. DeWinkel was awarded degrees in applied physics, civil and environmental engineering, and energy resources. He is currently on the staff of Wisconsin Power and Light Company.

Robin W. Gates is an energy analyst in the Wisconsin Office of Planning and Energy. He received degrees in environmental planning and resource management and in public administration.

Kathleen Briody Lipp holds degrees in zoology and environmental sciences. She is now a science research assistant to the Legislative Council of the Wisconsin State Legislature.

Mark Thornsjo is a marketing information specialist in energy conservation for Northern States Power Company. He graduated with a degree in urban and regional planning.

Stanley J. Kabala is director of the Pittsburgh Architects Workshop, which provides architectural and planning assistance for those unable to obtain it elsewhere. He holds a degree in public policy and administration.

ABOUT FRIENDS OF THE EARTH

We invite you to join Friends of the Earth, a dynamic environmental group. We work on the international scale to promote the most worthwhile option there is for the future—a living planet.

With national groups in 14 countries, active lobbies in Washington, DC and other capitals, and an aggressive publishing program of books and periodicals, we use every legal means at our disposal to promote changes that will allow Man to live in balance with this, our only planet.

All members receive *Not Man Apart* twice a month and are entitled to discounts on FOE books. Members in categories of $35/year and over receive FOE gift books as premiums.

Dues in FOE are not tax-deductible, in order that we may lobby vigorously.

Friends of the Earth Foundation solicits deductible contributions of money or property.

A membership application and book-order form appear on page 91.

RESOURCES FROM
FRIENDS OF THE EARTH

Soft Energy Paths
Toward a Durable Peace
By Amory B. Lovins, introduction by Barbara Ward

Soft Energy Paths crowns the critical work of Amory Lovins and FOE with a new program for energy sanity. This is an epoch-making book, the most important Friends of the Earth has published.

Lovin's soft path can take us around nuclear power, free industrial nations of dependence on unreliable sources of oil and the need to seek it in fragile environments, and at the same time enable modern societies to grow without damaging the earth or making their people less free.

Soft Energy Paths provides a conceptual and technical basis for more efficient energy use, the application of appropriate alternative technologies, and the clean and careful use of fossil fuels while soft technologies are put in place.

Harper & Row/Colophon edition
231 pages
$3.95

The Energy Controversy
Soft Path Questions and Answers
By Amory Lovins and his critics
Edited by Hugh Nash

In the last two years the theories of Amory Lovins have been condemned and defended; criticism has been led by advocates of the big, centralized energy systems that are now leading us into one energy crisis after another. The debate climaxed in joint hearings of two Senate committees, where Lovins refuted his major critics, point by point.

The Energy Controversy recaps those hearings, presenting Lovins and his opponents in dramatic dialogues on this most central issue of our time. How much does the soft path cost? What dangers lurk on our present hard road? How much would *it* cost? And how do we shift from one to the other?

Lovins is ". . . one of the Western world's most influential energy thinkers. His visionary essay, 'Energy Strategy: The Road Not Taken?', is described by population biologist Paul Ehrlich as 'the most influential single work on energy policy written in the last five years.' 'He has put a lot of things together that other people have been dealing with in pieces for years,' says John Holdren, an energy expert at the University of California at Berkeley. 'Because of him, the energy debate will never be the same.' Lovins's thesis has yet to be disproved, and his influence is widening daily. . . . The day after he talked with Lovins, President Carter told an international energy conference that the world should consider alternatives to nuclear power. . . ."

250 pages
Cloth: $12.50
Paperback: $6.95

—Newsweek

]88[

Frozen Fire:
Where Will It Happen Next?
By Lee Niedringhaus Davis

An LNG accident could be as bad as a reactor meltdown, and the major exporters of liquefied natural gas are OPEC countries.

But the gas industry wants to make it common, building terminals on most American coasts, Europe and Japan, bringing 125,000-cubic-meter shiploads of it in from the Middle East, Indonesia, and Alaska.

US trade in LNG is insignificant now, and world trade is small compared to industry plans. So there is still time to avert the dangers. If we do not, recent horrors with liquefied gases in Spain, Mexico, England, Abu Dhabi, and Staten Island may be the harbingers of a fearful future.

Lee Davis's book is the first major study of LNG prepared for the general reader. It tells everything you should know about the stuff before they try to put it in your town.

"The scope and logic of *Frozen Fire* should make it the Bible of anti-LNG groups throughout the world. It could even make converts of government agencies and the gas industry, preventing the inevitable LNG holocaust."

—Gene and Edwina Cosgriff
B.L.A.S.T.

"An impressive and powerful compendium of information on the dangers of another chemical threat to the so-called civilized world."

—John G. Fuller

256 pages
Cloth: $12.50
Paperback: $6.95

The Whale Manual
By Friends of the Earth Staff

The Whale Manual lays out the latest facts and figures about the great whales. Population estimates, habitats, and how quickly they are being killed, by whom, for what, and how—and what could be used as alternatives to whale products.

A special section outlines Friends of the Earth's controversial program to preserve the endangered Bowhead, with the help of the Eskimos who hunt it.

The Whale Manual is an invaluable source for readers committed to saving our planet's largest creatures.

168 pages
$4.95

Progress
As If Survival Mattered
Edited by Hugh Nash

The environmental book of 1978, *Progress As If Survival Mattered* presents Friends of the Earth's comprehensive and realistic program to take America away from energy crises and resource shortages . . . and charts the way to an abundant, healthy, and free future. This "Handbook for a Conserver Society" includes the thinking of our finest writers on energy, environment, and society.

"Here is a book undoubtedly headed for fame . . . an important anthology of long articles on issues which are critical . . . a guidebook for citizen action . . . an all-around introduction to the ills and successes of our planet, it should find its way into the homes of almost anyone who reads more than the Sunday comics. . . . Highly recommended."

—*Library Journal*

320 pages
$6.95

SUN!
A Handbook For the Solar Decade
Edited by Stephen Lyons

The Sun's rays inspired life on earth and stocked the planet with the fuel we've nearly exhausted. It's time we acknowledge the Sun's importance to our future.

By relying on the energy of the Sun, we can put an end to the costly and dangerous nuclear experiment. The solar future will be cleaner and freer, more equitable and enjoyable than the present.

Sun! is the official book of the International Sun Day movement.

"Here you have the thoughts of an extraordinary range of minds attending the solar nexus: Brower, Lovins, Bookchin, Illich, Goodman, Mumford, Grossman, Hayes, Long, Commoner, Georgescu-Roegen, von Arx, H. Odum, Inglis, Lyons, Stein, Shurcliff (especially), Reis, Goldman, Mills, Harding."

—*CoEvolution Quarterly*

". . . This is an overview of the social, economic, and technical advantages of solar energy. There is an extensive bibliography and a listing of organizations active in this field. Recommended. . . ."

—*Library Journal*

". . . a crusading, provocative, and prophetic work."

—*Los Angeles Times*

364 pages
$2.95

To: Friends of the Earth
124 Spear Street
San Francisco, CA 94105

Please Join Us.

☐ Please enroll me for one year in the category checked, entitling me to *Not Man Apart* and discounts on selected FOE books.
(*Contributions to FOE are not tax-deductible.*)

☐ Regular = $25   ☐ Spouse = add $5
☐ Supporting = $35*   ☐ Life = $1000***
☐ Contributing = $60**   ☐ Patron = $5000***
☐ Sponsor = $100**   ☐ Retired = $12
☐ Sustaining = $250**   ☐ Student/Low Income = $12

*Will receive free a paperback volume from our *Celebrating the Earth* Series.
**Will receive free a volume from our *Earth's Wild Places* Series.
***Will receive free a copy of *Headlands* (our award-winning, gallery-format book).
☐ Check here if you do not wish to receive your bonus book.

☐ Please accept my *deductible* contribution of $ _____ to Friends of the Earth Foundation (*checks must be made to FOE Foundation*).

Please send me the following FOE Books:

Number	Title, price (members' price)	Cost
_____	*Frozen Fire* at $12.50 ($10.00)	_____
_____	paperback *Frozen Fire* at $6.95 ($5.95)	_____
_____	*Energy Controversy* at $12.50 ($10.00)	_____
_____	paperback *Energy Controversy* at $6.95 ($5.95)	_____
_____	*Progress As If Survival Mattered* at $6.95 (5.75)	_____
_____	*SUN! A Handbook for the Solar Decade* at $2.95 ($2.25)	_____
_____	*Soft Energy Paths*, H&R ed. at $3.95 ($3.25)	_____
_____	*The Whale Manual* at $4.95 ($3.95)	_____
	Other FOE Titles:	

_____   _____
_____   _____
_____   _____

Subtotal _____
6% tax on Calif. delivery _____
Plus 5% for shipping/handling _____

☐ Send full FOE Books catalogue.   TOTAL _____
☐ VISA   ☐ Mastercharge
Number _____ Expiration date _____
Signature _____
Name _____
Address _____
City _____ State _____ Zip _____

]91[